WINNING AT MATH TRANSITION

Mathematics Study Skills Guide for Students Preparing for College

Paul D. Nolting, Ph.D.
Learning Specialist

Academic Success Press, Inc.
FLORIDA

This book is dedicated to my son, Eric, to my wife, Kim, and to the memory of my good friend Bill Thomas. It is also dedicated to the thousands of students who are having difficulty learning mathematics and the teachers who are teaching them.

Winning at Math Transition:
Mathematics Study Skills Guide for
Students Preparing for College
(First Edition)
by Paul D. Nolting, Ph.D.
First Edition Based on Winning at Math-Concise, 2018

ISBN 978-0-940287-72-3
Published by Academic Success Press, Inc.
Editor: Daniel Crown
Assistant Editor: Kimberly Nolting
Photo Credits: Unless otherwise noted, all photos and clipart are reproduced with the permission of
Fotosearch.com, and ClipArt.com.

Requests should be addressed to
Academic Success Press
3547 53rd Ave. W.
PMB 132
Bradenton, FL 34210
www.AcademicSuccess.com

Printed in the United States of America.

Contents

Preface

High School and college students have a complicated relationship with mathematics. Some love math's logic and challenges. Others dislike the subject and have difficulty learning the concepts. Complicating matters, students from both groups often deal with anxiety, procrastination, and failure. While students are allowed to feel any way they want about math, the necessity of the subject is unavoidable. Students must pass math to graduate from high school and college. While primarily aimed at helping students transition from high school to college math, this self-help guide is designed to assist all students in achieving their graduation goal, whether they are still in high school, home school, a GED track, or are returning to college after a prolonged absence.

There are three major obstacles to becoming successful in college mathematics. The first is that college math requires math-specific study skills that high school students do not often have. The second is that most high school students are only familiar with traditional lecture-based classrooms. College math is now taught primarily through computer-based applications, and the courses that *are* taught through lectures move four to five times faster than their high school equivalents. The third problem is that most high schools do not train their students to be independent learners. In modern college classrooms, students must acquire the skills necessary to learn math on their own.

It is sobering but true that these three issues are causing significant math learning problems. The average pass rate for math courses in college is about 50%. This guide is meant to help you wind up in the better half of this ratio, by helping you transition into college math having already learned the skills required to succeed in a redesigned classroom.

Fifty years of research exists on how math study skills and motivation affect success in mathematics. This research, which includes my own Ph.D. dissertation, indicates that these skills greatly increase students' ability to thrive in their courses. Study skills and motivation tactics—which we in the field often refer to as "affective characteristics"—account for up to 40 percent of a student's grade. These skills are extremely learnable and are applicable to college *and* high school students. That said, students with different learning backgrounds will encounter a variety of challenges in college math courses. Let's discuss some of these backgrounds and how we might address each of their unique challenges.

High School Students

Current high school students taking algebra and pre-calculus courses usually attend traditional lectures, which are spread out over an entire academic school year. Most of their learning occurs in the classroom and their grades are based upon test scores, homework, projects, and extra credit. In college, however, math grades are primarily based upon tests. Students cannot rely on homework or extra credit to keep their grades up. Additionally, college level courses require a "C" or better to advance to the next course. This means that high school students who plan to attend college should learn the study skills required to succeed at the next level as soon as possible. These students can use this guide to develop study-skills, test-taking techniques, and anxiety-reduction techniques that will serve them throughout their academic careers.

With this in mind, it is not too late for students who have already graduated. These students can take The Math Study Skills Evaluation - Concise, which will tell them which chapters and pages in this text will best serve their particular needs. After first reading these pages, they can finish the rest of the guide. I also strongly encourage high school graduates to practice the skills in this book alongside various online resources. YouTube tutoring videos are particularly useful, as many of them are produced in a compelling and entertaining fashion. Numerous phone applications also exist, which are likewise useful for students to practice specific math skills.

Home School Students

Home school students are in a similar situation as high school students, but, because they likely do not work directly with instructors, they are more likely to be used to computer programs or online math courses. These students, however, will also face unique challenges in modern college courses, as they must learn to work with tutors and other students in classrooms and math labs. To succeed in college it is important to know how to interact with instructors. It is also possible that these students are not accustomed to timed tests. These students need to practice math learning techniques, as well as anxiety-reduction and test-taking skills.

GED Students

Students obtain a GED for many different reasons. For GED students, math is usually one of their most difficult courses. Like home-schooled students, they are not used to lectures. The person aiding them may or may not have a math background. GED programs do not teach students math study skills or how to become a successful college student. This guide helps these students learn how to read college textbooks, take proper notes, reduce anxiety, and motivate themselves to succeed on tests. By practicing the skills taught in this book *while* finishing their GED coursework, students can increase their odds of smoothly transitioning into college math. Students who have already graduated with their GED can follow the same suggestions given above for traditional high school students. Whether reading this text before or during their first college math course, these students stand to benefit from the Math Study Skills Evaluation and what it tells them about how they learn.

Returning College Students

These students may not have completed their last college math class or maybe made a "C." Research shows that these students struggle in their next math course with up to a 75% failure rate. For students who did better in their first course, a few traps remain. Their next course might be taught in an entirely different design, which will require different skills and techniques to succeed.

If you struggled in your first math course, or if you are entering a different style of course, it is important to enter your next one with new tactics. Simply put, the methods you used in your last course did not work. And if they did not work the first time, then they will not work the next. The skills in this guide provide you with new strategies that can help you turn the corner on your collegiate math career.

Additional Resources

Winning at Math helps you become successful in mathematics by teaching you how to study on your own. This is true whether you are enrolled in a traditional math course, a modular course, or an emporium model course. To improve your chances for success, go to AcademicSuccess.com, click on the "Free Trial Math Study Skills Evaluation," and take the MSSE-C for a diagnostic evaluation of your math study skills. This evaluation does not contain math problems, but rather measures what you know about learning math. I also strongly encourage students to use the applications and websites listed in Appendix C. These tools have been proven to help with the transition between high school and college math.

Transitioning to College Math and Taking Control of Your Learning

1

In Chapter 1
You will learn these concepts:

✓ How to transition to college math

✓ How assessing math study skills improves study behaviors

✓ How studying math differs from studying other subjects

✓ How to get off to a quick start in math learning

✓ How to take control of math and motivate yourself

Transitioning to College Math and Developing a Productive Attitude

After hundreds of years of stability, the college math classroom is changing in new and at times intimidating ways. Whether you are entering your first college math course directly out of high school, after years of homeschooling, or after a prolonged absence, the environment you are about to enter is entirely unique to the new millennium.

While it is possible that your first college math course will involve a teacher standing in front of a dry-erase board, it is just as likely that you will take part in a modular or emporium model course, which involves one teacher overseeing a whole classroom of students who do their work exclusively on computers. It is also possible that you will wind up in a co-requisite course, which teaches multiple levels of math at the same time. Other courses are often accelerated, which means that you will have to complete the course in a mere eight weeks!

Universities and colleges now require students to become efficient independent learners. While you will often have either a teacher or a tutor around to answer questions, in most cases you will have to learn how to thrive and succeed on your own. To do this requires excellent math study skills, anxiety reduction techniques, test-taking skills, and motivational skills.

If you do not already have these skills, it is not your fault! For the most part, teachers don't teach this stuff in high school math courses. This book is meant to help you develop the techniques needed to succeed in a modern college setting. Based upon decades of research, the skills taught in this book are specifically designed to prepare you for the demanding nature of redesigned math classrooms. It does so by treating you like you are *already in* one of these courses. It lays out numerous tactics and strategies for everything from taking notes in computer-based classrooms to how to properly interact with your professors.

Before we get to any of this, however, it is important to first address your attitude toward math and how your relationship with the subject directly affects your success.

How to Develop a Productive Attitude

It is no secret math is not a popular subject these days. Not that this is anything new. From the moment man first began counting on his fingers, math has been the ultimate love or hate subject—even among the greatest minds in history. For every math-centric thinker like Pythagoras or Isaac Newton, there is at least one contrarian like Augustine of Hippo, a 5th-century philosopher who once claimed that mathematicians had made a covenant with the devil "to darken the spirit" and "confine man in the bonds of Hell."

While most modern math students may not take their disdain for the subject quite as far as jolly-old-Augustine, the fact remains that many view math differently than they do other courses. According to a recent poll conducted by Amplicate.com, nearly 86 percent of high school students claim to "hate" algebra. The same data shows that geometry is by far the most reviled subject for students under the age of eighteen—this despite the fact that many of these same students claim to enjoy school and look forward to attending their other classes.

If you are a college student, and the previous description rings a bell, you need to remember two things. First, there is nothing wrong with disliking math. Second, plenty of students who hate math pass it anyway. A negative math attitude is only dangerous if it becomes the basis for bad behavior. If a negative math attitude leads to poor class attendance, poor concentration and poor study skills, however, then you have more than a bad attitude—you have a problem.

The key to combating a bad math attitude is to develop a persistent attitude. Persistence, perhaps more than any other behavior, is absolutely crucial to success—not only in terms of passing math and graduating from college, but also in determining one's ability to thrive in the outside world. As former U.S. President Calvin Coolidge once wrote: "Nothing in this world can take the place of persistence. Talent will not; nothing is more common than unsuccessful people with talent. Genius will not; unrewarded genius is almost a proverb. Persistence and determination alone are omnipotent. The slogan 'press on' has solved and always will solve the problems of the human race."

In other words, even the smartest, most capable student cannot succeed in college without the ability to learn from failure and adversity. In the same way, those who struggle with school—or specifically math—are just as capable of success, so long as they "press on" through their problems.

By coupling a positive attitude with proper study and test-taking skills, you are capable of rising above any challenge college might throw at you—big or small. If you see math as one of these challenges, that is perfectly okay. The key is to remain persistent. Don't give up. Remember, the harder the challenge, the greater the reward.

Taking the Math Study Skills Evaluation —Concise

Math courses are not like other college courses. Most college courses only require you to read, understand and recall subject material to pass a class. To pass math, however, an extra step is needed; you must use the information you have learned to correctly solve math problems. This may sound like common sense, but this fourth step is a true game changer. It means that the general study skills you use in other classes are insufficient for math.

To get better grades in math, you need to learn math-specific study skills. These skills help you build confidence and achieve better grades. They also help you reduce anxiety—especially during your first college math course. This book not only covers these study skills, but also allows you to custom build your own study plan by addressing your unique strengths and challenges.

Before learning what study strategies work best for you in high school or college, it is first important to take the Math Study Skills Evaluation—Concise. To access the MSSE-C, visit AcademicSuccess.com— then click on "Free Trial Math Study Skills" and select "MSSE-C." This program determines what you know about studying for math and helps you determine what material in this book will help you the most. The MSSE-C involves answering a series of questions about your basic study behaviors. If it asks about behaviors you have yet to experience, click on what you think you might do in the scenario given. It does not contain any math equations or questions. Your answers result in an overall score between 35 and 100. When you finish, you will receive a results sheet, which presents information on the sub-categories you already understand, as well as those areas on which you can improve.

Your overall score and accompanying advice will look something like this:

> Sally College, the overall result of your evaluation is a score of 47. A score of 70 and below means you need to improve your math study skills and this could be the main reason you are having difficulty learning math.

You will also receive specific results on many areas, including chapters and page numbers to read:

> You have a score of 48 in Memory and Learning, which measures your understanding of learning styles and the learning process, as well as your ability to develop a useful learning plan. You need to read and study chapters 2 and 5.

If your results seem low, don't worry! This low score may explain your poor math grades. If you have spent your entire academic life using inefficient study skills, this is likely the source of your problems. The skills you'll read about in this book will directly address this.

Activity 1.1 Math Study Skills Evaluation—Concise Results

Student Name______________________________________ Date____________________

Advisor/Teacher Name____________________________ Email____________________

Overall result_____________
A score of 79 and below means you need to improve your math study skills, and this could be the main reason you have had difficulty learning math.

Study Effectively_____________, which measures how well you understand that studying for math differs from studying from other subjects. It also measures your effective use of study goals, reading/homework systems, study tools, and motivation. The reference chapter is Chapter One.

Memory and Learning_____________, which measures what you know about learning styles, the learning process, as well as developing a learning plan and memory strategies. The reference chapter is Chapter Two.

Reading and Homework_____________, which measures how well you understand a course's syllabus and what you know about developing reading and homework strategies to improve math learning. The reference chapters are Chapter Five and Chapter Six.

Classroom Learning_____________, which measures the ability to develop listening strategies, note-taking systems as well as the ability to ask questions. The reference chapter is Chapter Four.

Test Anxiety and Test Taking_____________, which measures how well you understand the effects of test anxiety, how to reduce test anxiety, how to take tests and how to analyze test results. The reference chapters are Chapter Three and Chapter Seven.

Based on your scores and a discussion with your advisor/teacher, list from most important to least important the order of the chapters in this text you most need to read:

Chapter___,
Chapter___,
Chapter___,
Chapter___,
Chapter___,
Chapter___,
Chapter___.

I will complete reading these chapters by this date: ________________________

I will also read the indicated pages: ________________________

How Learning Math Is Different

To learn and thrive in your math courses, it is important to understand the basic elements of the learning process. Math courses require you to do four things:

1. Understand material,
2. Process material,
3. Apply what you have learned to correctly solve a problem, and, finally,
4. Remember everything you have already learned to learn new material.

These four tasks help you master the math you must learn. If you do not complete one of the tasks, you will not complete the others. In other words, in order to successfully complete math problems, you must actively engage with the topics you are studying in real time. This is not true in most other courses.

Example: Political science courses require you to learn about politics and public service, but your instructor will not make you run for governor to pass the course. In math, however, you must be able to correctly solve problems. When you learn about factoring techniques, for example, you must solve numerous factoring problems in order to pass a test.

Math as Sequential Learning

Another way math differs from other subjects is that it uses a sequential learning pattern. The material you learn on one day is used the next day and the next day and so forth.

Example: Learning math is like building a house. Houses are built from the ground up: you build the foundation first, then the walls, then the roof. Math is the same way. It must be learned in a specific order. You cannot learn complex problems without first learning simple ones.

Unlike other subjects, you cannot forget important material after a math test. Each chapter in a math book is a foundation block for the next chapter. All building blocks must be included to successfully learn math. This building-block approach to learning math is the reason why it is difficult to catch up when you get behind. You can't skip information in one chapter to hurry up and get caught up in the next chapter. To succeed in math, each previous chapter has to be completely understood before continuing to the next chapter.

Sequential learning also affects studying for math tests. If you understand Chapter 1 and 2 but do not understand Chapter 3, you will not understand Chapter 4. When your test time approaches, you end up trying to teach yourself too many concepts while studying for the test when you should be reviewing and practicing. This makes it highly likely your results won't represent the best you can do.

Now that you understand how learning math is a building experience, what should you do if you have already fallen behind?

1. *Don't get anxious.* Stay calm.
2. *If your college* has a diagnostic math inventory in the tutoring center or math lab, take it to see what math concepts you have forgotten, and ask your instructor where you can go to relearn these math concepts.
3. *Take the time* to follow through. Many students give up too easily, or they think they will catch up a little at a time. Don't give up. The energy you put into your class at the beginning of the session will be more productive than energy put into class at the end of the session when you try to learn everything during the last week before the final exam.
4. *Study and really* learn the math; don't practice mimicking it.
5. *When it is time* to register for the next session, register immediately so you will be able to get into the math class you need. Why do all this? Because, again, math is sequential!

Math as a Sequential Course

Now that you understand the "building block" nature of math, think about your math history.

- What are your previous math grades?
- How well did you do on the math placement test at your college?
- How long has it been since you took a math course?
- When you look at your math history, are there times when you did not take math?

These questions are important because if you allow too much time to pass between different math courses, you may forget important concepts that you will need in a future class. In other words, your foundation might require a few repairs. This means that it is incredibly important to take a math placement test before enrolling for any college-level math course.

Math placement test scores are determined by how well you have learned previous math and whether the sequence in which you took your previous courses has helped you build a strong foundation. If you barely score high enough to be placed into a math course, then you will have gaps in your knowledge base. Learning problems then occur when new math material is based on one of these gaps.

In addition to considering how well you place into a course, the age of the placement test score must also be considered. Placement test scores are designed to measure your current math knowledge and are best used immediately.

Example: While a student can maintain success by taking U.S. History I and then waiting a year to take U.S. History II, taking Beginning Algebra and then taking Intermediate Algebra a year later decreases his or her course success rate by 20%, according to Miami Dade College research.

Other research indicates that even skipping one session greatly decreases math course success rates. The same problem has also been documented at Rutgers University where students were allowed to take their developmental courses whenever they wanted. Many of these students took developmental courses during their freshman years then waited to finish their math courses as seniors. This caused major problems as the students forgot everything they learned and still needed over a year of math courses to graduate. After conducting this research, the university implemented a policy requiring students to continuously enroll in their math courses to improve their course success rate.

College math courses should be taken in order, from the fall session to the spring session. If possible, avoid taking math from the spring to fall sessions. It is easy to forget important concepts over summer break. Summer math courses are also a good idea, unless the sessions are very short. Extended summer sessions, which typically last between 10 and 12 weeks, are preferable to waiting three months to take another course. If you cannot take summer math courses, make sure to review the last chapter in your last course to refresh your memory during your vacation. If you do not have your text, consider visiting math websites, such as www.purplemath.com. This will better prepare you to make a good grade in your next math course.

Math as a Thinking Subject

In a math class, you have to think from the minute the instructor begins talking. Note taking is challenging because you have to think about the math as your instructor explains a problem, all while writing the information down in your notes. In most cases, you are not just copying down facts. This could cause problems in note taking because you are concentrating on understanding the explanation while also trying to write down each problem step.

Also, students must think when reading the math textbook, which is why it takes longer than reading other texts. Remember when I said math is sequential?

Math is a series of steps and combinations of rules and properties. Students have to think between each step, remembering the rules and properties they have learned throughout their entire collegiate careers. When reading a math textbook, take time to think how each new concept connects with the previous one. This slows the reading process down significantly.

Since math is a thinking subject, what should you do?

1. *Reserve much more* time to read a chapter or complete your homework.
2. *During the evenings* or days when you don't have time to complete the entire homework assignment, at least review your notes and do a couple of problems to keep the information in your brain.
3. *You need to learn* how to think like a math instructor. This means you need to think in a step-by-step way to solve problems, writing down each step of the problem without skipping steps until the problem is solved.

Math as a Speed Subject

In most cases, math is taught faster than your other subjects. Math instructors have a certain amount of material to cover each session. They have to finish certain chapters because the next math course is based on the information taught in their courses. In many cases, math departments give a common final exam to gauge your readiness for your next course.

Another way math is a speed subject is that most math tests are timed. This causes many students to fear that they will not finish. THIS OFTEN CREATES PANIC! What makes me curious is, if students feel like they don't have enough time to complete the math test, why are most of them gone before the test time is over? Sure, students who have learned the math thoroughly may complete the test early. That makes sense. Some students leave, however, because they don't know the material, they want to leave the anxious environment, or they carelessly work through the test.

So, since speed is an issue in learning math, what should you do?

First, to use an analogy, start a daily workout program to stay in shape. Review, review, review as you learn and digest new material.

Second, practice doing problems within a time constraint. This prepares your for upcoming tests.

The Differences Between High School and College Math

Math, as a college-level course, is almost two to three times as difficult as high school-level math courses. There are many reasons for the increased difficulty: course class-time, the amount of material covered, the length of a course, and the grading system, among others. College math instruction for the fall and spring sessions is usually three hours per week; high school math instruction is usually provided five hours per week. What is learned in one year of high school math is learned in one session of college math. Simply put, in college, students receive less instructional time per week and instructors cover twice the material per course. As a result, most of the learning occurs outside of the college classroom.

During orientation sessions, I often ask students how much time they used to spend studying for math in high school. Without fail, most answer that they'd studied between one and five hours. Others tell me that they had paid attention in class, barely studied at all and still managed to pass. This method DOES NOT WORK in college. In order to succeed in a college-level math course, you must study at least 10 hours per week. This sounds like a lot, but when broken down into one or two hour study sessions, the time is not so intimidating.

Course-Grading System

While in high school, if you make a borderline D/F in a math course, your teacher might give you a "D," and you may continue to the next course. However, in some college math courses, a "D" is not considered a passing grade, or, if a "D" is made, the course will not count toward graduation. College instructors are more likely to give the grade of "N" (no grade), "W" (withdrawal), or "F" for barely knowing the material. This is because the instructors know students are unable to pass the next course if they have learned only a part of the curriculum. Also, in most high schools, you may graduate by passing one to three math courses. Some college programs require four math courses, and students must make at least a "C" in all of them to graduate.

What should you not expect in a college math course? First, since math is not as subjective as English, do not expect to talk a math instructor into extra work to earn a better grade. Second, in college, there are no "daily work" grades—though you still have to do your homework. Don't expect to do well without doing the homework even though it is not collected. Third, test scores may be the only grades that count toward your final grade. Therefore, do not assume you will be able to "make up" for a bad test score.

Putting Math Into Perspective

One way to change the way you learn math is to compare it to one of your favorite hobbies or extracurricular activities. The following activities all relate to math in fundamental ways.

Math as a Puzzle

Have you ever tried to put a puzzle together with one or more of the pieces missing? It becomes very frustrating, especially if you think all the parts are there and, after hours and hours of work, you realize you won't be able to finish. Puzzles are fickle that way. Each piece is crucial. If even a single piece is missing, you are left with an incomplete picture. This is also true when you are halfway through solving a math problem and find out you do not know or have forgotten the rules for the next step. In order to solve math problems, you need to understand all the rules. In other subjects, you may forget some of the rules and still pass a test. In math, however, not knowing or forgetting one major concept often causes you to miss problems.

Math as a Foreign Language

Another way to properly study math is to treat it as a foreign language. Like a foreign language, math has unfamiliar vocabulary words, which must be put in sentences called expressions or equations. Understanding and solving a math equation is similar to speaking and understanding a sentence in a foreign language. When read or spoken aloud, mathematical symbols are translated into words and sentences. Math sentences use symbols (spoken words), such as:

"=" (for which you say, "equals"),
"-" (for which you say, "less"), and
"a" (for which you say, "unknown").

If you do not understand these symbols, you are in serious trouble. This is best illustrated by the experiences of a student named Charlie, whom I once worked with. While preparing for a math test, Charlie memorized how to work every type of problem his teacher had covered in the previous weeks. Despite his efforts, he still failed. When I went over the test with him, I asked him how to factor a polynomial. He asked me, "What is a polynomial?" I then asked him what a monomial was, and he did not know. I talked to his tutor and the rest of the tutoring session was spent learning the language of mathematics by developing vocabulary notecards so he could understand what we were talking about. He had the ability to do math but could not understand the language. Learning how to speak math as a language is key to success. With this in mind, to stay fluent in a foreign language, you need to use it every day. Math is no different. If you want to retain your ability to speak and read mathematics, you must practice every single day.

So, now that you know math is like a foreign language, what should you do?

1. *Start a vocabulary* list in the back of your notebook. Include the definitions and examples the instructor uses in class. Writing this vocabulary list helps you study too!
2. *Preview the chapter* before class in order to identify vocabulary words that you will hear in class. You don't have to understand all of them while previewing the chapter, but familiarize yourself with the words. This will help you improve taking notes in class.
3. *Practice saying* the vocabulary words out loud. Get comfortable with the words, their meanings, and how they connect with one another.

Math as a Sport

Learning math is similar to learning to play a sport such as basketball, track or football. In order to find success, you must actively practice. You can listen and watch your coach all day, but unless you practice those skills yourself, you will not learn and probably won't even get into the game or meet.

For example, in basketball, the way to improve your free-throw percentage is to watch and understand the correct shooting form and then to actively practice the shots. If you simply listen to your coaches describe the correct form and watch them demonstrate it, but you do not practice the correct form yourself, you will never improve.

Math as a Musical Instrument

Learning math is like playing a musical instrument. To master a musical instrument, you need to understand music theory and learn the various hand-eye movements required to strike the right sound and tone.

This involves lots of practice. You can see someone play the piano, cello or electric guitar, watch their hands and hear the sounds, but unless you practice, you will not learn how to play. Imagine how many concerts you have attended or seen on television, or how many people you have watched play the piano. Could you now go play the piano, cello or electric guitar? No. You need to practice these instruments before playing them well just like you have to practice math before you learn it.

Many of your other college courses can be learned by methods other than practicing. In social studies, for example, you can learn information by listening to your instructor, taking good notes and participating in class discussions. Many first-year college students mistakenly believe they can approach math the same way. Ask an experienced math student, and you will hear a different story. Face it, unless you are a math genius, you must practice. So, how do you practice math?

- *First,* make sure you understand the math concepts and the basics.
- *Second,* practice the homework problems over and over until the answers come to you naturally.
- *Third,* learn the vocabulary associated with problems.
- *Fourth,* try to "teach" the math to someone else.

Activity 1.2 Studying Math is like...

One of the best ways to study for math is to treat it the same way you would one of your favorite hobbies or performance-based activities. Not to say that you need to enjoy math the same way you enjoy playing basketball or singing or playing an instrument, but you certainly need to give it the same type of attention. Answer the questions below to find out how you can use your experiences participating in an activity you love to shape or reshape your approach to math.

1. Which one of the comparisons in this section is the most relatable to you? Do you play a musical instrument? Have you ever taken a foreign language class? Have you ever played a sport? Choose one of the comparisons and explain how you can use it to improve your study habits.

2. If none of the comparisons given in this section apply to you, think of one of your hobbies or activities and explain below how you can use your practice routine for this activity as an example for how to properly study math.

3. Explain, in your own words, why it is important to consistently study for math. What, for instance, makes studying for math unique?

Quick Start to Learning Math: Lecture, Emporium, and Modular

Now that we have put learning math into proper persepective, it is time to discuss individual methods that work in each of the different types of modern classrooms. With this in mind, this section provides a Quick Start Guide for Lecture, Emporium, and Modular classrooms. Math is a speed subject. As in an Olympic sprint, students must got off the blocks early so they do not fall behind. We will expand on the strategies listed on the next few pages in the chapters to come. In the meantime, however, familiarize yourself with this content so that you can get off to the best start possible in your new math class.

Tradtional Lecture Courses

For years, most students attended standard lecture courses, where an instructor stood in front of the classroom, wrote problems on the board, and discussed steps. Afterward, the instructor would ask if students had any questions. While these classes still exist, many colleges are replacing them with accelerated, contextualized, or co-requisite designs. Accelerated courses finish by midterm, contextualized courses teach two courses at the same time and co-requisite courses have a mandatory additional course or lab. These courses all require students to speed up their math learning. This means that it is now crucial to become a faster and more efficient learner. Here are a few tips for regular lecture and accelerated courses.

1. *Attend Class.* When taking an accelerated classroom course, it is vital to attend all classes. It is a good idea to make a study buddy, so that if you do miss a class he or she can give you that day's notes. Also, either ask your instructor if additional resources exist that summarize the material you missed, or ask if your teacher offers the same class at another time. If he or she does, attend this second session. Finally, go to the math lab for assistance in learning missed material. You cannot afford to get behind.

2. *Listen Intently.* Accelerated math courses go four to five times as fast as high school courses. This means that it is important to pay close attention. To listen more intently, sit in an area of the classroom that has a direct view of the instructor. These areas usually have the least distractions. This might be front and center in the classroom or somewhere in the middle back. Listen for phrases that let you know the instructor is serious about what he or she is saying—phrases like, "students often get stuck on this step"; "a lot of people miss this problem"; or, "this is tricky."

3. *Take Good Notes.* If necessary, ask your instructor if you can record lectures with your smartphone (usually just audio). It is important to develop a note-taking system. At the top of every page, jot down the date with the chapter and section number. Make sure you take note of the vocabulary words your instructor uses the most. For now, use the left side of your notebook for key words and the center of the page to record problem steps. Make sure to leave spaces between the steps you don't understand. This way you can fill them in later. If you get confused, ask your instructor if you can take a picture of a problem on the board. This way, you can take it to a tutor. It is also smart to review your notes with a study buddy.

4. *Always Complete Your Homework.* Use your smartphone to set up a homework schedule. Commit to when and where you plan to study, as well as the number of homework hours you need per week (at least 8-10). The best time to do homework is right after class (preferably in a math lab). It is important to establish a homework system. If you are doing your homework from a textbook, then you must record each problem and each step in a separate homework notebook. This way you can revisit these problems later. For online homework, make sure to take notes on the problems on which you use the built-in help system. Again, make sure to record each step. If you have problem with an assignment, seek out useful Apps or websites. Go to the math lab for any remaining unsolved problems.

5. *Develop a Reliable Test-Taking System.* If Review your notes and homework problems before each test. Seek out practice tests from a math lab or from your instructor. Complete these practice tests several days before your real test. If you cannot find a practice test, them make one. If you have test anxiety, practice the relief methods listed in Chapter Three. It also helps to stick to a reliable test-taking method. Luckily, we provide such a method in Chapter Seven. Before taking a test, it is important to analyze your past behaviors. This way you can choose strategies that will keep you from leaving problems blank or leaving the classroom early out of frustration.

Learning in an Emporium or Modular Math Course

Learning math in an Emporium Model, modular or computer assisted course requires enhancements to lecture-based math courses. These enhancements can be added to or sometimes replace strategies. (Author: Rochelle Beatty, Ph.D).

Listening, Viewing, and Note-Taking Skills. It is important that notes be taken and kept in a binder dedicated to the course, even when most of the materials and assessments are archived or completed digitally. If watching videos is a lecture component of learning, notes should be taken. During video viewing, pay close attention to any images or titles that are added to the screen.

Look for places where the presenter pauses or looks at the camera. These are clues as to the importance of material just presented or coming up and should prompt you to take notes. If the video contains closed captioning or a transcript that can be printed, this information needs to be put in your notes.

Often the lecture component of the instruction also contains power-points that need to be viewed. During note-taking of the power-point material, pay attention to slide headings, bold print, bullet points, material contained in a box, worked out examples or vocabulary words. The style used when presenting the material provides a hint with regard to the importance of the information given. If printing files is an option, print a copy of the slides and add those to your notebook.

Don't forget to look at the book when completing a math course that has a major computer component. Make sure you know where to find the eBook in the digital materials. The eBook, typical format for this type of course design, is often rich in multimedia content including videos, animations, tutorials and additional electronic examples. These are provided in pace with the concepts covered in the text so they create a richer learning environment.

Completing Your Homework. To organize notes on multi-media provided within the assigned homework set, use a homework grid similar to what is used for note-taking. See an example below.

The first column will be used for lecture notes on material that is included in the problem set. Then the next two columns are used to show your work on each problem. Pay attention to the problem numbers that may be listed in the heading of homework problems.

This often shows an area in the text that is related to that problem. Although your homework may contain some media components, don't assume these are the only ones available. Look in the eBook section correlated to your problems, or places in the digital materials, to know all media types that are available to assist you in learning. Also remember the instructor is a great resource.

Homework systems are programmed with multiple attempts. Always seek out assistance from the tutorials provided, the instructor, a student aid, or a tutor before using your last attempt. If someone is not readily available, skip the problem and go onto the next one. Make a note on your homework problem sheet so you know to go back and get help on that one before you submit your work. Make sure your work is neat so you can study it for a test. Don't solely rely on the internet for studying. Have some of your own materials to use for review.

A schedule is probably provided that suggests deadlines for completing homework, quizzes and exams. Always stay ahead of the published class schedule. Life is full of curve balls. By working ahead, time is left for those unexpected life moments requiring a shift in our schedule.

Notes from video, power-points and media within the assignment	Problem #1	Problem #4
	Problem #2	Problem #5
	Problem #3	Problem #6

Studying For and Taking Tests. Check to see if a chapter or unit review exercise set is available as part of the digital course materials. This exercise will provide additional practice for and test question clues. To review, check digital course material to see if you can take a practice test with the help button turned off. Take the test under real timed conditions to see if you have enough time to finish the test and check your answers. Make sure to reserve enough time in your schedule to complete the review. You should complete the review at least two days ahead of time to give you enough time to get help for solving the problems you could not get right. As you are completing the review, create a 5X8 notecard with key ideas and information needed to successfully complete the test and questions.

It may not be an option to use this note card on the test, but this is still an important exercise to complete. Remember just writing the notecards can help you remember information for the test. Go back and review your notes and comments from the media portion of assignments and add that information to the notecards. Make sure the information on the card is organized in "bite-size" pieces so you can review it quickly. If time permits, review the note card often during the week of the exam. The notecards can be studied right before the test for additional confidence by memorizing formulas and principles.

Check to see if there are any study groups or study sections made available for the tests. Studying with others and in a group can have benefits. You may want to consider asking to Skype with fellow classmates as a study group. Even a study group of two can help. Also remember to follow the pre-test guidelines given in this text.

When taking tests, write down the problem and show all of your work. Often, instructors award partial credit. But if they don't have anything to see, there is not an opportunity to award additional points. Don't rely on only the points provided by the computer. Your instructor is an important part of your course assessment. Have an archive of your work so he or she can be of assistance in your success.

Using Technology. Technology can be unreliable, but that is not an excuse for falling behind in your work. In the front cover of your course notebook, archive a few pieces of important information.

Write down the number of the tech support for the technology being used in your course. There may be times where there is an issue with the program and you need to call to get assistance.

Write down at least two locations where computers are available on campus. You may need to use a campus computer at some point in the course. Know where to find these resources and there times of operation. Write down two places of business close to your home where you can use the internet for free (a public library, a coffee shop). We rely on having strong internet connections; however there are times when internet connections can be interrupted. Have a back-up plan when presented with this dilemma.

Summary

The Quick Start section in this chapter is intended to help students put math study skills to immediate use in lecture, Emporium Model, and modular classrooms. Put the skills listed above to use right away, then build upon them as you complete each chapter in this text. For your convenience, the list below details where you can find specific skills.

Chapter One: Attending Class

Chapter Two: Learning Strengths

Chapter Three: Managing Anxiety

Chapter Four: Listening and Note-Taking

Chapter Five: Time Management and Reading

Chapter Six: Homework

Chapter Seven: Test-Taking

How to Take Control of Your High School or College Math Experience

The best ways to take control over math are to develop an internal locus of control, to avoid learned helplessness, and to eliminate (or at least decrease) procrastination. This is first accomplished by setting short-term goals. Accomplishing these goals allows you to see the relationship between positive learning behavior and mathematics success. You can then develop long-term goals such as making an "A" or "B" in your math class or graduating college.

Defining Locus of Control and Self-Efficacy

Locus of control has to do with the locus, or location, in which a student places the control over his or her life; or, in other words, who or what a student feels controls his or her behavior and grades. Students who feel that conditions beyond their control prevent them from getting good grades have an external locus of control. These students blame instructors, home conditions and money problems for their poor grades. In essence, external students feel their lives are controlled by outside forces, such as fate or the power of other people. They often believe that these problems are beyond their control.

Other students feel that they have the power to control their situation, and this power comes from within. These internal students take responsibility for their success, while most external students reject responsibility. Internal students believe that they can overcome most situations, since results depend on their behavior or personal characteristics. Internal students accept responsibility for their behavior and realize that studying today will help them pass the math test scheduled for next week. Internal students can delay immediate rewards. Instead of going to a party, they study for tomorrow's test. Internal students also know when to postpone putting in extra hours at their workplace in order to study for a math test.

Generally, locus of control means that students who are internal will work harder to meet their educational goals than will external students. The internal student can relate today's behavior (e.g., studying, textbook reading) to obtaining a college degree and gainful employment.

On the other hand, external students cannot connect the behavior of studying today with getting passing grades and obtaining future career opportunities. Thus, internal students are more oriented toward making high math grades than are externals.

Self-Efficacy

Self-efficacy is another term colleges and universites often use. This term is related to locus of control but is more specific to certain tasks. Locus of control is an overall belief that you can be successful. Self-efficacy is more specific to the belief that you have the abilities or skills to complete certain tasks (Bandura, 1982).

To determine self-efficacy, ask yourself this question by filling in the blank, "Do I have the skills and ability to succeed at ______." The blank could include tasks such as employment, an English course, history course, math course or graduating from college.

Just like in locus of control, education research has found out that self-efficacy is an important success predictor of student motivation and self-management behavior (Schunk, 1991). Students with high efficacy are more likely than students with low efficacy to complete difficult tasks, to put in more effort, to show persistence, to have used different learning strategies, and to have less fear about completing a task. In short, these students believe that they can complete tasks such as taking notes, reading the textbook, completing their homework, and managing their time. They persist at these skills until proficient.

Before reading this text, you may not have had these skills and may have suffered from low self-efficacy. Now, you can use these learning and test-taking skills to improve your self-efficacy in math, which will lead to more motivation to become successful in future courses.

Productive Persistence

Productive persistence is the newest term associated with math success. The Carnegie Foundation defines productive persistence as the tenacity and good strategies students need to use to become successful. According to the Carnegie Foundation's research, the combination of tenacity and good study strategies is a formula for math success.

Tenacious students work hard, study the appropriate amount of hours, make learning a priority, and postpone immediate self-gratification for long-term success in math. These student remain in math classes and do not withdraw at the first sign of difficulty. They also develop short-term and long-term goals. This allows them to see the progress they make even when they hit a stumbling block. You can improve your tenacity by applying the motivational and goal setting suggestions in this chapter.

The good strategies part of this successful formula involves learning and using math study skills. Those students who use general study skills often study 15 to 20 hours a week on math without any success. I call these students "wheel spinners." By not learning math-specific study skills, these students are in effect riding a stationary bike. They are putting in a whole lot of work, but they aren't going anywhere.

The math study skills you will learn in this text have been proven to improve math success. Using these skills as part of your My Math Success Plan allows you to fulfill the Carnegie Foundation's formula.

Self-Learning

Math courses require students to learn a great deal of information outside of class. Luckily, modern math students have more tools at their disposal than ever before. Over the course of this text, we will introduce you to numerous supplementary tools that will help you learn complex mathematical concepts. Some of these resources are physical locations on your campus, others are Web-based research tools.

How to Improve Your Self-Esteem

Many students experience problems in both their academic and personal lives because they lack self-esteem. Self-esteem is the part of our personality which allows us to feel good about ourselves and enjoy our accomplishments. Having self-esteem means that you respect others and have a sense of peace within yourself. Students who have self-esteem also have a "can do" attitude about accomplishing their goals.

Students with poor self-esteem may not put forth as much effort to accomplish their goals. These students can improve their self-esteem by taking responsibility for their feelings, thoughts, abilities and behaviors. Students with poor self-esteem need to change their negative emotional reactions into positive emotional reactions.

Improving self-esteem can be accomplished by changing the negative emotional statements you sometimes tell yourself and by changing bad behaviors that result in a lack of self respect.

By developing positive short-term goals, you change the behaviors that are associated with poor self-esteem. Negative emotional statements are often referred to as "negative self-talk." It is possible to change these negative statements to positive ones. List your negative self talk statements starting with "I" and change them to positive "I" statements. For more on how to accomplish this, see activity 1.3.

10 Ways to Improve Your Self-Esteem

Improving your self-esteem is an important step to improving your success in mathematics. Use the following positive statements to replace any negative thoughts you have during math class.

1. When I do well at something, I am going to congratulate myself.
2. I am going to stop procrastinating and blaming others for my problems.
3. When I fail at something, I am not going to blame myself but find out how to be more successful the next time.
4. I will not worry about what others think of me.
5. I will do something I like to do at least one day per week.
6. I will keep a "To Do" list, and I will feel good about myself when marking off each completed item.
7. I will set up short-term and long-term personal/educational goals.
8. I will like myself and have the courage to take risks to change.
9. I will ask for help without feeling guilty.
10. I will put five positive statements about myself on a 3x5 card, and I will read them to myself when I feel bad.

Mindfulness and Success

From the very first week of college orientation until the hectic week of graduation, college life is full of opportunities and responsibilities. It does not matter if you live on campus or commute. It does not matter if you are single, married, with or without parental responsibilities. Balancing opportunities and responsibilities is key to academic success. That is why just about every college orientation textbook has a chapter on values, goals and time management. One ability successful students use continuously is the ability to manage these opportunities and responsibilities so that they do not become distractions. There is another powerful source for distractions ---- our minds.

The mind can be our worst enemy or our closest ally. Thoughts, emotions and sensations about previous experiences or future ones can serve as our looking glass into the present. As the worst enemy, a present situation triggers old thoughts and emotions that cause us to handle the present in ways that we further harm ourselves. As our closest ally, when emotions and thoughts about what is happening around us disrupt our ability to focus, the mind can learn to stop these internal distractions and improve concentration in the present.

One of the most successful strategies is practicing the mental state called Mindfulness. Mindfulness is the ability to interact with the present moment with full attention, without allowing thoughts, internal dialogue, and accompanying emotions to interfere with how we interact in the present.

Sounds great, even somewhat simplistic, but it takes time and the desire to work with the anxiety instead of avoiding it. It means a student with high math anxiety must live with it. So, instead of delving into the "how to's," we need to explore the benefits of Mindfulness.

1. Anxiety minimizes the ability to learn. Anxious intrusive thoughts while completing math problems decrease speed and accuracy (Brunye, Mahone, et. 2013) Students with too much anxiety cannot remain focused while studying, so they waste precious time. Mindfulness consists of strategies to remove the source of anxiety and thus the anxious emotion from the present moment.
2. For many students, math anxiety is embarrassing, thus they try to hide it. A key concept in Mindfulness is that anxiety is the same as any other emotion. Anxiety, like all other emotions, does not define who you are unless you let it. When learning to develop a state of Mindfulness, people learn to accept anxiety as an emotion while also learning to love themselves even with it.
3. Mindfulness is not limited to math anxiety. A mental state of mindfulness enhances how we interact in any present situation, so students with math anxiety who make the effort to learn Mindfulness will improve other areas of their lives.

You can do the following steps to achieving this dimension of Mindfulness immediately.

Step 1: Practice the Right Mindset.

- Non-judgement: Emotions are not good or bad. We can't judge ourselves based on the emotions we feel, even though others may.
- We are not our emotions: Emotions come and go. We are like the sky and emotions are the clouds. Sometimes there are dark gray storm clouds and other times white fluffy ones. None of them stay in the sky forever. When we feel nervous or anxious, we tell ourselves that it is not going to last.
- Avoidance does not work in this situation. We can't keep ourselves from feeling emotions. We must learn how to interact with them. It is almost like personifying the emotion and setting it to the side, then deciding what we are going to do with it.
- Self-respect and self-love: We cannot allow our minds to beat us up. There are enough tangible stresses to manage. We do not want to battle our emotions too. We must learn to accept our emotions, in this case anxiety, and manage it in ways that it does not interfere with studying.

Step 2: Learn how to react to anxiety when it first creeps into the mind:

- When you notice it, say, "I am beginning to feel some anxiety." Whether you know why or do not, say, "I need to deal with this issue and the anxiety, but for now I must set it aside."
- The following ideas seem touchy feely, but they can work. Take a piece of paper, wad it up, and as you throw it in the waste basket, say, "Not now. Not going to happen. I have time to deal with it later." Each time anxiety or pre-occupation of any kind that is interfering, throw another piece of paper in the waste basket. You can do this before a math class, studying alone or in a lab. When you take a math test, throw the paper wad into the basket before the test.
- Breathe deeply several times and as you inhale think of a positive outcome for completing the study task. Keep this thought inside you as you hold your breath for a few seconds. Then exhale and relax your body.

Step 3: Use interactive study strategies while studying

- Create test review notecards while learning new math concepts, rules, and formulas. Each card should include definition, how it is used, steps, and example(s). Keep these cards in your room. Take pictures of each card with your phone, so when you have a few extra minutes, you can review them.
- Try studying in a math center, study center, library or other places where you can concentrate but not be alone. The mind is less likely to begin on a destructive path when you are not alone.
- Find someone to study with a few times a week. If you live off campus, use the many sources of technology to connect and study with others in your class.

Step 4: Get some sun and exercise.

- It only takes fifteen to twenty minutes of sun each day to lift the spirits. If possible, let the sun hit your arms or legs so that the body can absorb vitamin D. This could happen just by walking to class, walking the dog, or playing with your kids outside.

Step 5: Do not participate in avoidance

- Emotions come and go, but that does not mean we should ignore them. Anxiety comes in different forms and levels of intensity. If anxiety is controlling you right now, do not wait to read Chapter Three on math and test anxiety. Read it now. Also, seek out professional assistance, so an expert can coach you throughout the semester on how to minimize the anxiety and manage it.

Activity 1. 3 Taking Control of Your Math Experience

Math students have various perceived levels of math control. Even the best students need to develop strategies to improve motivation and control over math. This is first accomplished by setting up short-term goals. To get started follow the directions below:

1. List and describe how you would use one locus of control concept and one self-efficacy concept for motivation and controlling your math experience

    a.

    b.

2. List and describe how you would use two productive persistence strategies to control your math experience.

    a

    b

3. List and describe how you would improve your self-esteem by changing negative math self-talk statements into positive statement to improve motivation. First list the negative statement and then list the positive self statement to take its place.

    Negative statement:

    Positive statement:

    Negative statement:

    Positive statement:

4. Describe what mindfulness means to you. Then list five mindful thoughts/behaviors that can help control your anxiety and help you control math. Ex: " I will not let my mind beat me up about math:

Mindfulness to me is:

Thoughts/behaviors:

a.

b.

c.

d.

e.

How Learning Math Pays Off

Now that we have learned how math is different, how to end negative self-talk, and why your first math test is so important, what do we want to gain from learning math?

Another way to ask this question is, "What are your reasons for attending college?" Are you going to college to better yourself? Are you going to college to obtain the career you want? Are you going to college to make more money? I ask these questions to many students who are taking math, and what do you think is the most popular response? You're right! Most students say they are going to college to make more money. Students also indicate they want to like their careers and make enough money to live a better life. It does not matter if these students are going for an Associate of Science degree, an Associate of Arts degree or a Baccalaureate degree. Their goals are to graduate in the shortest amount of time and to obtain a high paying career.

Students also often indicate they want their career to be secure so they will not have to keep switching employment or have to be retrained. This is especially true of men and women who are returning to college to obtain better careers to support their families. Being clear on your college goal is a way to motivate your college success. So does taking math courses mean you can obtain a higher paying career, have more career options, and put together a competitive resume? YES!

Associate of Science degrees are usually two-year degrees once you have been admitted to a program or once you have completed the prerequisite courses. Four of the five highest paying jobs a student can get with an Associate's degree (dental hygienists, diagnostic medical sonographers, registered nurses, radiation therapists) require college algebra or statistics. The national salary for these careers range from about $60,000 to $80,000 with register nurse being the lowest and computer occupations being the highest paid career. Four of the five lowest paying jobs (veterinary technician, medical records technician, physical therapist assistant, and occupational therapist assistant) require mathematics courses below college algebra and statistics. The national salary for these careers range from about $25,000 to about $55,000 with veterinary technician being the lowest and radiological technologist being the highest paid position. Based on this information, if you want the most job selections and highest paying career, then passing more mathematics courses is your ticket to success.

Students who are going for a Baccalaureate degree either obtain an Associate of Arts degree at a community/ junior college or are already attending a university. In most cases, these students will make more money than students with Associate in Science degrees. One way to look at these majors is to plot the graph by using salaries and the increasing amounts of math courses. The following graph demonstrates the 20 top careers with the best pay, fastest growth, and the most openings for the 21st century. It also shows national annual salary per year (Farr and Ludden, Best Jobs of the 21st Century, 2013).

Look at the pattern. All five highest paying jobs (software developer, network administrator, construction manager, engineer, computer systems analysts) require business calculus or higher. In fact, all nine careers that have salaries around $50,000 a year require business calculus or a higher math course. Four of the five lowest paying jobs (writers/authors, film editors, music arrangers, graphic designers) require mathematics courses below college algebra and statistics.

By now, using your excellent number sense, you have figured it out. The more math courses you take, the higher paying career you will get! Learning mathematics is the key to finding success in your personal, academic, and professional lives.

Now that you know that taking more mathematics means more career choices and better pay, how can you accomplish this goal? Effective learning strategies for successful math study skills leads to more math success. Math success helps you graduate and allows you to select majors in areas that lead to high paying jobs and more job security.

Obstacles to Success

Sometimes, self-talk and statements by others may block your motivation to become successful in your math course. Don't get in the habit of saying, "When will I ever use this math?" or "Why do I have to take this math course because I will not use it in my job?"

These statements may make you feel better when you are not successful in a math course; however, they also lead to poor motivation and less studying. The real questions are: "Do I need this math course to graduate?" or "Do I need this math course as a prerequisite to enter my major?"

Don't let obstacles get in the way of your success in math. Stay persistent and obtain the career you've always wanted.

For example, if you are a business major, in most cases you will need applied calculus to be accepted to the college of business. Being successful in applied calculus is your "ticket" to your business major and making that million dollars.

In some cases, you may have to take developmental math courses in order to be ready to take required math courses. Don't be discouraged! I have personally worked with hundreds of students who began in the lowest developmental math courses and now have their business, engineer, computer science or economics degrees. Thousands of students have finished their developmental math courses and have completed college algebra or calculus and are now in their careers. Remember, more math equals more money.

Best Jobs Requiring a Bachelor's Degree

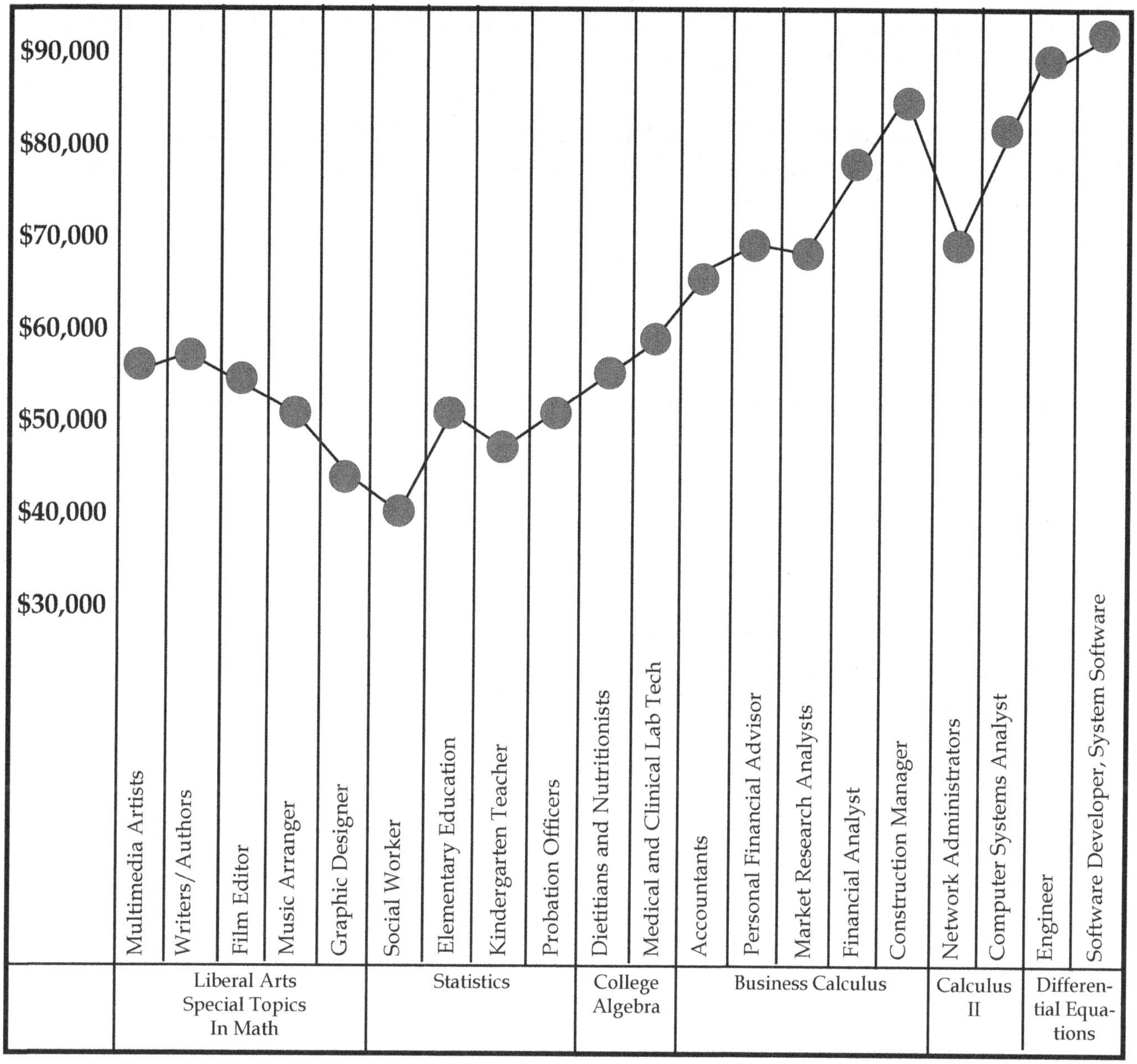

Source: *Best Jobs for the 21st Century*, Sixth Edition © JIST Works 2012
Graph: © Academic Success Press Inc. 2013

Activity 1.4 What type of career do you want?

Choosing a career is tricky. Sometimes our interests and our desire for a highly compensated career do not match up. This means that many students who do not like math will still have to take numerous math courses to obtain their degree. Fill out the form below to find out just how much math you will need to take to graduate and start your career.

1. What type of career do you want to have when you finish college? If undecided, then write down the general field you might want to go into (humanities, science, business, etc.)

2. How much money do you want to make in your career? Write the total in the space below.

3. Look at the chart on the previous page. Do you see your career? If so, see if your career goals match the amount of money you want to make. After you have done this, write down the math courses you will need to take to finish your degree.

4. If your chosen career is not on the chart, either use the Internet to look up the information you need, or visit your school's career or guidance counselor. Follow the same procedure as the question above.

Chapter One: Checklist for Knowledge and Behaviors

Put a check after understanding the knowledge statements and fill in the behavioral activities.

Chapter Knowledge

1. I understand that persistence is one of the key factors to math success. ______
2. I understand that math requires sequential learning, which means that once concept builds on the next concept. I must build a firm foundation with each chapter to help me succeed. _____
3. I understand that one of the major differences between high school and college math is that most learning occurs outside the college classroom and I need to study more. _____
4. I think that learning math is like learning a foreign language or playing sports or a musical instrument can help me change my math study skills. _____
5. I understand that math is an active subject and I must practice it by doing my homework. ___
6. I understand that the skills required for learning math differ from the skills required from learning other subjects. ______
7. I understand that passing most courses requires reading, understanding and recalling the information however math must generalize concepts by thinking like their instructor. ______
8. I understand that math is the only socially expectable course students can hate and fail. However, I know that I can dislike math and pass it at the same time. _____
9. I will go to math class every class day. _____
10. I will use the quick start strategies for my lector, Emporium or modular course. ____
11. I visited the instructor the first or second week of class. ________
12. I developed motivational strategies for myself. ______

Behavioral Activities

1. I completed and reviewed the Math Study Skills Evaluation-C with a total score of ____.
2. I know that my last math grade was a _______ and have not taken math in _______ months. I believe I have a good math foundation or one that needs some repairs (circle one).
3. I studied _________ hours a week for math in high school and will study ______ hours a week for math in college.
4. I know math is not just numbers so I started a math vocabulary list in the back of my note book on / / .
5. I found out from the career center that my major will need ______ math course and make $________.

Name: ______________________________________ Date: ______________

Assignment for Chapter 1

Answer the following questions on a separate piece of paper.

1. Why is math considered to have a sequential learning pattern?

2. Give two examples of how math is a speed subject.

3. Give three examples of how college math is different from high school math.

4. Select the quick start design (Lecture, Emporium or Modular) that you are taking, then list and describe five learning strategies you can immediately begin this week.

5. Define Productive Persistence and how you obtain this characteristic.

6. Define "self- esteem," then list and define five ways you can improve your self-esteem.

7. Define "mindfulness" and list three mindfulness strategies you can use to improve math success.

8. How can positive and negative attitudes toward math affect learning?

9. Describe three personal obstacles that may prevent you from learning more math.

10. List one strategy per obstacle (listed above) that you will use to overcome it.

Assessing and Using Your Math Learning Strengths

2

In Chapter 2 You will learn these concepts:

✓ How math knowledge affects math success

✓ How a student's affective characteristics influence success

✓ How to determine an use your best learning style

✓ General memory techniques

✓ How to develop "My Math Success Plan"

Ingredients for Success in Learning Math

After exploring the characteristics of math, the second step to improving your ability to study the subject is to understand your unique study skills. Just as a mechanic does a diagnostic test on a car before repairing it, you need to complete a diagnostic test to identify which of your go-to study strategies need improvement. This allows you to focus only on those learning areas that most benefit you. There is no sense in wasting time on areas in which you are already sufficiently proficient.

In order to do this, you need to develop a Math Success Plan. At the end of this chapter, you will write down your various strengths and challenges. This list will eventually become an overall attack plan for your future math courses.

Before we start exploring your learning skills, however, it helps to understand the keys to academic success. Dr. Benjamin Bloom, a famous researcher of educational learning, discovered that IQ (intelligence) and "cognitive entry skills" (knowledge about math) account for 50 percent of a student's course grade. This is highlighted in the Bloom Chart, printed on the next page. As you can see, quality of instruction represents 25 percent of a course grade, while "affective student characteristics" reflect the remaining 25 percent. New research suggests that these characteristics account for 41 percent of academic achievment. (Zientek, et. 2013)

Dr. Bloom's four variables to academic achievement are as follows:

- *Intelligence,* for our purpose, is considered to be how fast a person can learn or relearn math concepts of varying difficulty.
- *Cognitive entry skills* refer to how much math people already know before entering a math course.
- *Quality of instruction* concerns the effectiveness of math instructors, lab assistants and tutors when presenting material in the classroom, math lab and while tutoring. This effectiveness depends on textbooks, curriculum, teaching style, tutoring style, teaching aids (videos, apps, websites, online homework) and other assistances.
- *Affective student characteristics* are characteristics people possess, which affect their course grades, excluding how much math they knew before entering a course. These affective characteristics include anxiety, study skills, attitudes, self-concepts, motivation, test-taking skills, locus of control, and productive persistence.

What You Know About Math Affects Your Grades

Going into a class with poor math knowledge sets you up for achieving low grades. A student placed in a math course that requires a more extensive math background than he or she possesses will probably fail that course. Without the correct math background, you may fall behind and never catch up. The following information is important to know in order to make sure that you are placed in the proper math course.

Figure 1
Variables Contributing to Student Academic Achievement

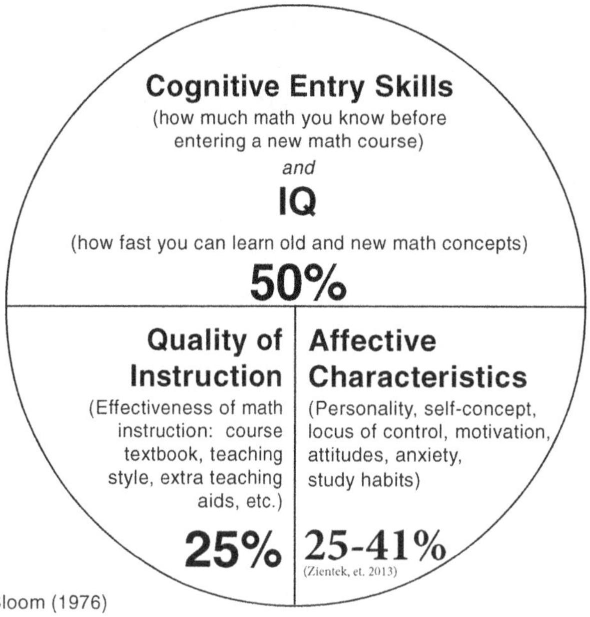

*Bloom (1976)

Placement Tests and Previous Math Grades

The math you need to know to enroll in a particular math course can be measured by a placement test. Most colleges and universities use a standardized test such as the ACT, SAT, or Accuplacer to place students into math courses. However, some institutions use their own tests to place students. It is important to remember that the grades you earned in prerequisite math courses measure your level of math knowledge. However, a few students are still incorrectly placed in math courses by these resources. So, it is important to talk to your advisor and teacher to provide information that might assist in properly placing you into a math course. Students in high school or college who made a "C" or "D" in their last algebra course are likely to fail their next algebra course.

If your placement scores are questionable or borderline between two classes, ask the advisor if there is another diagnostic test you can take. You can also try speaking with a math instructor because he or she may be able to ask questions that will help determine correct placement into a math course. An inaccurate evaluation of math ability and knowledge can only lead to frustration, anxiety and failure; a proper placement will set you up for success!

What if you think you are not ready for the course in which you have already enrolled? Without the correct math background, you might fall behind and never catch up. Most students do not want to add another math course to their lives, but taking a course that is too difficult will most likely result in problems. You will either withdraw and then re-enroll next session or, worse, fail and then take it again. So, observe what is taking place in your math class and determine what is best for you! Be honest! You want to be successful! Remember the building blocks in Chapter One? You want to build a firm foundation when enrolling in math courses. Getting into the right course is a major factor in building that foundation.

How Quality of Instruction Affects Your Grades

Quality of instruction accounts for another 25 percent of your grade. This instruction includes such things as classroom atmosphere, the instructor's teaching style, lab instruction, textbook content, and format. All of these "quality" factors affect the ability to learn in the classroom.

Interestingly, probably the most important "quality" variable is the compatibility of an instructor's teaching style with the students'.

First, you need to discover your learning style. Second, try to find an instructor who best matches your learning style. Sometimes this is difficult to find out, but a good place to start is to talk to the chair of the math department. The department chair knows the faculty. Advisors, too, know the teaching styles of every member of a math faculty. The key is to start your inquiry early enough to get into the class that bests suits your needs.

If you cannot find an instructor to match your learning style, improving math study skills and using a math lab or learning resource center can compensate for this. You and your instructor might not be the best match, but it is still your responsibility to be successful in the course.

The quality of tutors is also a major part of the effectiveness of a math lab or learning resource center. Trained tutors are essential for instruction because they know how to help students become better learners. Otherwise, tutoring sessions are nothing more than glorified study halls with a few helpers. The best way to work with tutors is to not only ask for assistance with math concepts but also to inquire about good study strategies.

Math departments design courses so that each course covers certain math concepts. Ideally, each course curriculum prepares students for the next level of math. However, sometimes the courses have gaps between them, or an instructor doesn't cover all the material required, leaving students unprepared for their next courses. If this happens, go to your instructor immediately to explain that you did not cover the math in your previous course. Your instructor will suggest a way for you to learn it.

How Affective Student Characteristics Influence Your Grades

Affective student characteristics account for about 25 percent of your grade. These characteristics include math study skills, test anxiety, motivation, locus of control, learning style, and other variables that determine your ability to learn math.

Most students do not have this 25 percent of their grade working in their favor (or 41 percent based upon Zientek). In fact, most students have never been taught any study skills at all, let alone math-specific study skills. It is also true that many students do not know their best learning styles, which means they may study ineffectively by using those that are least effective. Until recently, little attention has been devoted to how students learn, so it is not unusual if you do not know what your learning styles are. Later on in this chapter you will have the opportunity to explore the ways you learn most effectively.

By improving your affective characteristics, you benefit from more productive learning experiences and, subsequently, higher grades. Thousands of students have improved their math study skills, lowered their anxiety, and taken control of learning by using the strategies suggested in this book. The first step in improving how you study math is to determine your best math learning style.

Learning Style and Memory

Now that you understand the variables that contribute to success in mathematics, it is time to briefly discuss one final element: memory. While the ability to recall and apply learned knowledge is crucial to all fields of study, math is particularly demanding in this regard. As mentioned in Chapter One, math is a sequential subject. This means that math students must implant key concepts and ideas into their long-term memory for future use.

With this in mind, the rest of this chapter is designed to help you improve your ability to properly learn and remember math processes. In order to do this, you must first determine your preferred learning style. This way you will be able to choose the study skills presented in this chapter that will best suit your individual needs.

Determining Your Best Learning Style

A learning style is a description of the cognitive, affective, and physiological factors that shape the way a student inputs material to be learned and then demonstrates the knowledge of the material. Learning styles also pertain to the best time of day to study, environmental factors that may improve learning (silence, music, lights) and how the brain best processes the material. Research has shown that students who understand learning styles improve learning effectiveness in and outside of the classroom. Many learning style inventories are available, and students should talk to their instructors or counselors about taking one or more.

There are different types of learning styles assessments. One type focuses on learning modalities while other assessments focus on cognitive or environmental learning styles. In this chapter we will focus on modality learning styles.

Learning Modalities (Using Your Senses)

Learning modalities focus on the best way your brain receives information; that is, learning *visually* (seeing), *auditorially* (hearing) or *kinesthetically* (touching, hands-on). If one is available, take a learning style inventory that measures learning modalities. Even better, take an inventory that measures learning modalities specifically for math, such as the Learning Styles Inventory (Brown & Cooper, 1978) or the Learning Modality Inventory for Math Students (Nolting, 2008). Other learning-style inventories measure learning style, but they mainly focus on English or reading-learning modes. Sometimes students have different learning styles for math, so an inventory that measures subject-specific learning styles is ideal. Take the Learning Modality Inventory for Math Students, found in Appendix A, to better understand your math-learning style. Based on the Learning Modality Inventory for Math Students, rank order your learning modality score by putting 1, 2 or 3 in the space below. Number one is for the learning style that is most like you. If you have a tie, pick the style you feel is most accurate.

Auditory ______________

Visual ________________

Kinesthetic _____________

Modality learning styles are neither good nor bad. They are concerned with how you best take in information. Depending on the subject, professors teach in various ways. In the case of mathematics, most math instructors are visual learners, and they tend to teach the way they learn best. Most of these instructors will write on the board or use a PowerPoint presentation to help you learn math.

Using Your Learning Style to Improve Memory

There are many different techniques that can help you store information in your long-term memory and reasoning. Using your learning sense or learning style and decreasing distraction while studying are very efficient ways to learn. Using your best learning sense (what educators call your "predominate learning modality") improves how well you learn and enhances the transfer of knowledge into long-term memory/reasoning.

As you'll remember, the learning senses are vision, hearing, touching, etc. Ask yourself if you learn best by watching (vision), listening (hearing), or touching (feeling).

Another helpful tool is the Learning Styles Modality Inventory for Math Students, available in Appendix A. It is the only learning style inventory that has specific modality math learning style information. This means by taking the inventory, you learn the specific ways you prefer to learn math. Based on your preferred learning style, practice those learning suggestions first. Remember, learning styles are neither good nor bad. They are based on genetics and environment, not ability or intelligence.

The next few sections in the textbook cover the three major types of learners and provide specific tips on how each type can streamline the memory process. Find the section that best suits your particular needs and progress accordingly.

In the meantime, here is a quick refresher on five types of learners:

- *Visual Numeric Learners* prefer to learn by reading and writing.
- *Auditory Numeric Learners* prefer to learn through lectures and other verbal communication.
- *Tactile Concrete Learners* prefer to learn through hands-on learning.
- *Social Individual Learners* prefer to learn either by themselves or by one-on-one tutoring.
- *Social Group Learners* prefer to take part in group learning.

Multiple Senses

If you have difficulty learning material from one sense, try learning material through two or three senses. Involving two or more senses improves your learning and remembering. Review the figures in this section on using your learning styles. Whenever possible, combine learning styles. If your primary sense is visual, and your secondary sense is auditory, write down equations while saying them out loud. Writing and reciting the material at the same time combines visual, auditory, and some tactile/concrete styles of learning. Likewise, studying with a pen or highlighter is a visual as well as a tactile/concrete way to improve your concentration. Placing the pen or highlighter in your hand forces you to concentrate on what you are reading. After you write and recite the material back to yourself, do it five or ten more times to over-learn it.

Improving Memory for Visual and Auditory Learners

Most college students are either visual or auditory learners. While it is easier to find math teachers and tutors to accommodate these learning styles than it is for other methods, students who learn this way should still learn to maximize their chances of success. With that in mind, read the section on this page that is specific to the way you prefer to learn and then take a look at the adjoining chart (printed on the facing page). These charts present easy-to-implement strategies that are custom designed to help specific types of learners improve their success in math.

Visual (Watching) Learner

Visual learners—also known as Visual Numerical Learners—study best by repeatedly reading and writing down material. These learners get more out of reading a textbook and watching their professors write on the board than they do listening to lectures.

A visual way to decrease distractions is by using the "my mind is full" concept. Imagine that your mind is completely filled with thoughts of learning math, and other distracting thoughts cannot enter. Your mind has one-way input and output, which only responds to thinking about math when you are doing homework or studying.

Auditory (Hearing) Learner

If you are an auditory learner—one who learns best by hearing the information—then learning formulas is best accomplished by repeating them back to yourself, or recording and listening to them. In-class and Web-based lectures are perfectly suited to your learning style, which gives you plenty of study options.

Reading out loud is one of the best auditory ways to get important information into long-term memory. Stating facts and ideas out loud improves your ability to think and remember. If you cannot recite out loud, recite the material to yourself, emphasizing the key words.

An auditory way to improve your concentration is by becoming aware of your distractions and telling yourself to concentrate. If you are in a location where talking out loud will cause a disturbance, mouth the words "start concentrating" as you say them in your mind. This usually increases the length of time you are able to concentrate.

Visual Numeric Learner

These students learn math best by seeing it written. If you are a visual numerical learner, you may learn best by following these suggestions:

1. Use worksheets, workbooks, handouts, additional math texts and any other additional written materials.
2. Play games with, and get involved in activities with, visual printed materials such as multiplication or algebra flash cards.
3. Use visually orientated computer programs, DVDs, homework programs, math websites, and phone applications like those mentioned in this text.
4. Watch YouTube videos on learning math.
5. Rework your notes using the suggestions in this text.
6. Make "3 X 5" note or flash cards putting the variables and numbers in different colors.
7. Use Study Stacks to develop your own virtual flash cards or use the virtual flash cards already developed (www.academicsuccess.com—Student Resources—Student Math Practice and Learning Sites.)
8. Use different colors of ink to emphasize different parts of each math formula.
9. Visualize numbers and formulas in detail.
10. Ask your tutor to show you how to do the problems instead of telling you how to do the problems.
11. Take pictures of the board with your cellphone, or use your phone to record a lecture.

Auditory Numeric Learner

If you are an auditory numerical learner, you may learn best by following these suggestions:

1. Say the numbers to yourself or move your lips as you read the problems.
2. Record your class and play it back while reading your notes.
3. Read aloud any written explanations.
4. Make sure all important facts are spoken aloud with auditory repetition.
5. Read math problems aloud and try solutions verbally as you talk yourself through .
6. Record directions to difficult math problems and refer to them when solving those specific types of problems.
7. Record math laws and rules in your own words, by chapters, and listen to them every other day (auditory highlighting).
8. Use your smartphone or tablet to record a lecture.
9. Explain to the tutor how to work the math problems.
10. Explain to group members how to solve math problems.
11. During the test, sub-vocally talk yourself through the problems.
12. Take the test in a private room and talk to yourself out loud to solve the problem.

Improving Memory for Tactile/ Concrete Learners

A tactile/concrete learner needs to feel and touch material to learn it. Tactile concrete learners, who are also called "kinesthetic" learners, tend to learn best when they concretely manipulate information. Unfortunately, most math instructors do not use this learning sense. As a result, students who depend heavily upon feeling and touching for learning will usually have the most difficulty developing effective math learning techniques. This learning style creates a problem with math learning because math is more abstract than concrete. Also, most math instructors are visual abstract learners and have difficulty teaching math tactilely. Ask for the math instructors and tutors who give the most practical examples and who may even "act out" the math problems.

As mentioned before, a tactile concrete learner will probably learn most efficiently by hands-on learning. Also, learning is most effective when physical involvement with manipulation is combined with sight and sound. For example, as you trace the face you also say the words out loud.

Based on the Learning Styles Inventory for Math Students, tactile concrete learners best learn math by manipulating the information that is to be taught. If you are a tactile concrete learner, you may learn best by following the suggestions in the figure printed on the next page. Try as many of these suggestions as possible and select and practice the best suggestions that help. If you do not have these manipulatives or don't know how to use them, ask the math lab supervisor or instructor if they have any manipulative materials or models. If your math lab does not have any manipulative materials, ask for help to develop your own.

Tactile/concrete learners can also use graphing calculators to improve their learning. Entering keystrokes makes it easier to remember how to solve the problems. This is also an excellent way to remember how to solve a problem when using a calculator during a test. Another way tactile/concrete learners can learn is to trace the graph with their fingers when it appears on the calculator. They should say and trace every equation to "feel" how the graph changes when using different equations.

A tactile/concrete way to improve your study concentration is by counting the number of distractions for each study session. Place a sheet of paper by your book when doing homework. When you catch yourself not concentrating put the letter "C" on the sheet of paper. This will remind you to concentrate and get back to work. After each study period, count up the number of "C's" and watch the number decrease.

Tactile/Concrete Learner

These students learn math best by hands on learning. If you are a tactile concrete learner, you may learn best by following these suggestions:

1. Cut up a paper plate to represent a fraction of a whole.
2. Fold up a piece of paper several times and cut along the fold marks to represent a fraction of a whole.
3. In order to understand math concepts, ask to be shown how to use Cuesinaire or algebra titles as manipulatives.
4. Try to use your hands and body to "act out" a solution. For example, you may "become" the car in a rate-and-distance word problem.
5. Obtain diagrams, objects or manipulatives and incorporate activities such as drawing and writing into your study time. You may also enhance your learning by doing some type of physical activity such as walking.
6. Find a way to physically interact with common mathematics techniques. For an example, follow the directions below.

F (a) (c)

O (a) (d)

I (b) (c)

L (b) (d)

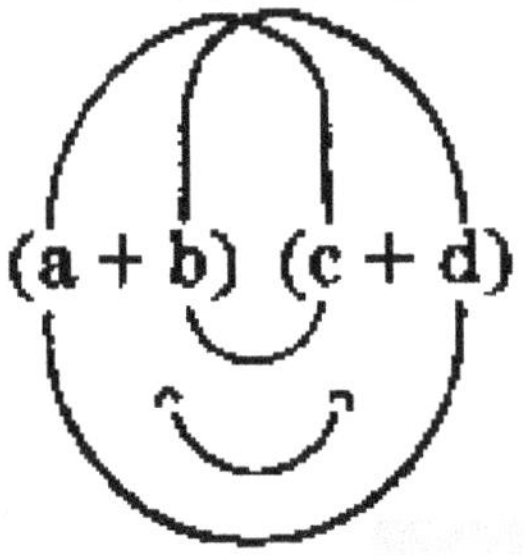

FOIL is used to remember the procedure to multiply two binomials. To use FOIL, multiply the following:

- the First terms ((a) (c))
- the Outside Terms ((a) (d))
- the Inside Terms ((b) (c))
- the Last terms ((b) (d)).

To learn FOIL, trace your finger along the FOIL route.

7. Ask to use the Hands-On-Equations Learning System using manipulatives to learn basic algebra. You can go to their website (www.borenson.com) to learn more about this system and other systems to help you learn math.
8. Go to one of the "learning stores," usually in your local mall, to see if they sell manipulatives. You can also try a K-12 learning resource center to see if they have manipulatives, such as magnetic boards, that you can put letters and numbers on and move around. Also, talk to the coordinator of students with disabilities to see if they use manipulatives when tutoring their students with learning disabilities.
9. Tear up a piece of paper into several pieces and put an x on some of the pieces. Mark the other pieces with numbers 0 to 9. The pieces with the x can represent the variable and the other pieces can represent the numbers. You can now use the pieces of paper to set up and solve equations.
10. Use the virtual manipulative websites at www.academicsuccess.com – Student Resource website or Google "college math manipulative."

Activity 2.1 Use Your Learning Style to Improve Your Math Knowledge

Now that you know your preferred learning style, it is time to put it to use. Use the questionnaire-printed below to figure out how to choose the resources that best fit your best way of learning.

1. Rank order your preferred learning style 1-3.

Auditory ____________
Visual ____________
Kinesthetic ____________

2. Think back to the last math course you completed. How did you study for this class? Did your approach follow your best learning style? Or did you adapt to the way your professor taught the course?

____

____

____

____.

3. The best way to use your learning style is to use your top two learning preferences at the same time to learn material. List three learning strategies that can combine your best learning preferences:

a.

b.

c.

4. Based on this learning style, what type of professor should you seek out? Keep in mind how this professor may or may not teach his or her classes.

____

____

____

____.

General Memory Techniques

Now that you know how to use your best learning style to improve your memory, it is time to learn a few techniques that help you remember specific information. These techniques are especially helpful when studying for tests.

A Good Study/Math Attitude

Having a positive attitude about studying will help you concentrate and improve your retention. This means you need to have at least a neutral math attitude (you neither like nor dislike it), and you should reserve the right to actually learn to like math. View studying as an opportunity to learn rather than as an unpleasant task. Tell yourself that you can learn the material and that learning it will help you pass the course and graduate.

Be a Selective Learner

Being selective in your math learning will improve your memory. Prioritize the materials you are studying. Decide which facts you need to know and which ones you can ignore. Narrow down information into laws and principles that can be generalized. Learn the laws and principles 100 percent.

Also, you must learn the math vocabulary in each chapter to continue to understand the instructor and math material.

Example: If you have been given a list of math principles and laws to learn for a test, put each one on an index card. As you go through them, create two piles: an "I already know this" pile and an "I don't know this" pile. Then, study only the "I don't know this" pile until it is completely memorized and understood.

Become an Organizer

Organizing math material into idea/fact clusters helps you learn and memorize it. Grouping similar material in a problem or calculator log are examples of categorizing information. Do not learn isolated facts; always connect them to other similar material.

Use Visual Imagery

Using mental pictures or diagrams is especially helpful for visual learners and those who are right-hemisphere dominant. Mental pictures and diagrams involve 100 percent of your brainpower. Picture the steps to solve difficult math problems in your mind.

Example: Use the Foil Method to visually learn how to multiply binomials. Memorize the face until you can sketch it from memory. During a test, you can then sketch the face onto your scratch paper and refer to it.

Make Associations

Association learning can help you remember better. Find a link between new facts and some well-established old facts and study them together. The recalling of old facts will help you remember the new ones and strengthen a mental connection between the two. Make up your own associations to remember math properties and laws.

Example: When learning the commutative property, remember that the word "commutative" sounds like the word "community." A community is made up of different types of people who could be labeled as an "a" group and a "b" group. However, in a community of "a" people and "b" people, it does not matter if we count the "a" people first or the "b" people first; we still have the same total number of people in the community. Thus, a+b=b+a. When learning the distributive law of multiplication over addition, such as a(b+c), remember that "distributive" sounds like "distributor," which is associated with giving out a product. The distributor "a" is giving its products to "b" and "c.

Use Mnemonic Devices

The use of mnemonic devices is another way to help you remember. Mnemonic devices are easily remembered words, phrases or rhymes associated with difficult-to-remember principles or facts. Chances are you've used these devices since elementary school.

Example: A mnemonic device to remember the Order of Operations is "Please Excuse My Dear Aunt Sally." The first letter in each of the words represents the math function to be completed from the first to the last. Thus, the Order of Operations is Parentheses (Please), Exponents (Excuse), Multiplication (My), Division (Dear), Addition (Aunt), and Subtraction (Sally).

Using mnemonic devices can improve a student's mathematics learning. Students making up their own mnemonic devices can remember them better than the ones given to them. Try to make up your own mnemonic device, but if you have difficulty, use the mathematic mnemonic devices on the Winning at Math Student Resource website.

Example: FOIL is a common math acronym. FOIL is used to remember the procedure to multiply two binomials. Each letter in the word FOIL represents a math operation. FOIL stands for First, Outside, Inside and Last, as it applies to multiplying two binomials such as (2x+3)(x+7). The First product is 2x (in the first expression) and x (in the second expression). The Outside product is 2x (in the first expression) and 7 (in the second expression). The Inside product is 3 (in the first expression) and x (in the second expression). The Last product is 3 (in the first expression) and 7 (in the second expression). This results in F ((2x)(x)) + O ((2x)(7)) + I ((3)(x)) + L ((3)(7)). Do the multiplication to get $2x^2 + 14x + 3x + 21$, which adds up to $2x^2 + 17x + 21$.

Use Acronyms

Acronyms are another memory device to help you learn math. Acronyms are word forms created from the first letters of a series of words. Using acronyms improves a student's mathematics learning. Making up your own acronym devices is the best way to remember them. If you are having difficulty making up acronyms then use the Winning at Math Student Resource website.

How to Use Number Sense

Number sense is a lot like common sense. It is the ability to see if your answer makes sense without using algorithms. (Algorithms are the sequential math steps used to solve problems.) An example of using number sense or estimating is in "rounding off."

Taking the time to estimate the answer to a math problem is a good way to check your answer.

Example: Solve 48 + 48 by rounding off. Rounding off means mentally changing the number (up or down) to make it more manageable to you, without using algorithms. By rounding off, 48 becomes 50 (easier to work with). 50 + 50 = 100. If the choices for answers were 104, 100, 98 and 96, you would then subtract four from the 100 (since each number was rounded up by 2) and you get 96.

Another way to use number sense is to check your answer to see if it is reasonable. Many students forget this important step and get the answer wrong. This is especially true of word or story problems.

Examples: When solving a rate-and-distance problem, use your common sense to realize that one car cannot go 500 miles per hour to catch the other car. However, the car could go 50 miles per hour.

The same common-sense rule applies to age-word problems where the age of a person cannot be 150 years but could be 15.

Further, in solving equations, x is usually a number that is less than 20. When you solve a problem for x and get 50, then this isn't reasonable, and you should re-check your calculations.

Also remember, when dealing with an equation, to make sure that you put the answer back into the equation to see if one side of the equation equals the other. If the two sides are not equal, you have the wrong answer.

If you have extra time left over after you have completed a test, you should check answers using this method. Always use all of your alloted test time.

Metacognition — Putting It All Together

Metacognition is a new concept in mathematics that looks at what students are thinking when solving math problems. Using your memory process to learn how to solve math problems is the first step, but you must also be able to apply concepts to solve homework or test problems. The key to your success is the self-monitoring of problem solving, not memorizing how to do the problems. Memorization of steps instead of understanding the rules and principles leads to passive learning and unsuccessful problem solving. It is important that you develop a reliable method. Gray (1991) in her article, "Ideas in Practice: Metacognition and Mathematical Problem Solving," suggests a math solving model based on metacognition. This model of plan, monitor and evaluate is a framework for solving math problems also supported by other researchers.

Planning consists of understanding what the problem wants, the strategies to solve the problem and potential obstacles. It also includes understanding what information is required, doing the calculations and predicting the outcome.

Monitoring is putting the steps in order, keeping one's place, identifying and finding errors, understanding when additional information is needed, knowing when to use another strategy and knowing when you have part of the answer.

Evaluating includes knowing if the answer seems right (number sense), putting the answer back into the equation, doing the opposite of the function to see if the answer is correct, and measuring the efficiency of the plan and monitoring.

Using metacognition to solve math problems increases your homework and test-taking success.

For more, see Chapter Six (How to Improve Your Homework Techniques) and Chapter Seven (How to Improve Your Math Test-Taking Skills).

Activity 2.2 Developing Your Memory Techniques

Now that you have learned about different memory techniques to help you remember math information, you can now develop your own. Follow the instructions below to develop different techniques.

1. Describe how you would use a good study/math attitude to improve learning.

2. Give an example of a formula you need to learn and how you would use visual imagery to learn it (cannot use book example)

3. Give an example of a math concept you are now learning and how you would use a mnemonic device or acronym to remember it (cannot use book example).

4. Explain number sense and give an example of how you would use it during homework and on a test.

   Number sense is:

   Two examples are:

Creating an Individual High School or College Success Plan for Math

In order to make the best grades in your high school or college math courses, you need to develop a detailed plan for every semester. This plan must include a list of goals, and the steps and resources required to meet these goals. This book is designed to help you produce a plan that meets all of these requirements and more.

To illustrate what your finished plan might look like, we've printed a completed copy of the form after this section. Notice that the student who filled out this form has written her goals in a clear, concise fashion. As you can see, to properly complete this plan, you first need to determine your best learning style. Next, you need to set your goals for your current math course(s). Finally, figure out what study strategies are best suited to your specific learning style, then list those strategies that address your semester goals.

To fill out your My Math Success Plan, see the blank form in Appendix B.

Why This Success Plan Is Necessary

Creating a math success plan for every semester is important for many reasons. Here are just a few of them:

- *Plans breed confidence.* Feeling prepared for a test or a class gives you the confidence you need to succeed.
- *It lets your professors know you are serious.* By filling out a math success plan, it lets your professors know you are making their class a priority.
- *It helps you figure out what does and doesn't work.* Making a success plan allows you to take stock in what study skills actually work for you.

Sections A and B

Section A is fairly straightforward. Simply list your name and year in college. List your current math course in Section B, and you are ready to move on!

Section C: Learning Information

Section C compiles everything you've worked on in this chapter—specifically your MSSE scores, your areas for improvement and your personal learning style preference.

Section D

Section D is where you should write down your semester goals. You may need a second meeting with your counselor to establish the rest of your goals. Remember, these goals should relate to your expected grades, study skills improvement, study schedule and, if applicable, services offered by the disability support service office.

These goals should include the following:

1. Your overall GPA
2. The grade for your math class
3. How to improve math study skills weaknesses from the Math Study Skills Evaluation
4. Deciding how many hours a week to study math
5. Deciding which days to study math
6. How many tutoring hours a week
7. How many times you will visit your teacher
8. To review each returned test with a tutor or teacher

Section E

Section E is perhaps the most important section of this activity. It is where you put together everything you have learned in this book. If you need to work on your note-taking skills, use this space to write down your intention to use the "Seven Steps to Math Note-Taking." If you need to work on your test-taking skills, write down what you have learned in Chapter 7.

What you write should be completely personalized. No two students share the exact same strengths and weaknesses. Be truthful and accurate.

As you complete this section, remember that you do not need to go into too much detail. Just write down the exact strategies you are going to use from this text, and make sure you remember where to find them. In fact, it is a good idea to write down page numbers next to each strategy.

Section F

Section F is where you log all of the motivational strategies you have learned about in this chapter. For example, if you have a problem with procrastination, you would write something like, "I will figure out the reasons for my procrastination and overcome them." This is also where you list the positive statements you have created to replace any negative talk.

Completing Your Plan

If you need help finishing your plan, visit any or all of the following:

- *a guidance counselor*
- *a favorite teacher*
- *a math teacher*
- *an ROTC advisor*
- *a parent*

All of these professionals are able to give useful advice on a wide range of subjects. Also keep in mind that you should revist this plan every time you take a test. If you do not make the grade you want, you might need to revise your plan. Seek help from your instructor in making these revisions. Ask for additional suggestions. If you *did* make the grade you wanted, you should share this information with the anyone with whom you have discussed your plan. Anyone who has helped you complete your plan deserves feedback, as it can reinforce efforts to devise and use effective strategies.

In the end, you will be glad to have completed your math success plan, as it will not only keep you on task during the semester, but also help you reduce anxiety in your future courses. Once you have a plan in place, it is much easier to maintain confidence (see example on next page).

My Math Success Plan (Example)

Semester: Fall 2014

A: Student Information

Name: Sally Student
Year in College: Sophomore

B: Course: Elementary Algebra/Algebra II

C: Learning Information

I am predominately a visual learner.

Based on my MSSE scores, I need to work on note-taking and homework techniques.

D: Semester Goals

1. I will obtain a 3.0 GPA.
2. I will make a "B" in elementary algebra/algebra II.
3. I will improve my math study skills weaknesses by working on the suggestions from the computerized Math Study Skills Evaluation.
4. I will set up a study schedule and each week will complete a study-goals sheet.
5. I will attend tutorial sessions.
6. I will see my math instructor every two weeks and after each major test.

E: Math Study Strategies

Use math note-taking system
Color code notes
Develop a math glossary
Use online homework system
Learn relaxation techniques
Develop my own test-taking system

F: Motivation Strategies

1. I will make an appointment with my math teacher.
2. I will reward myself when completing a short-term goal.
3. I will learn the reasons for my procrastination and overcome them.
4. I will tell myself positive statements everyday.
5. I will tell myself by passing math, I will get my business degree.

Chapter Two: Checklist for Knowledge and Behaviors

Put a check after understanding the knowledge statements and fill in the behavioral activities.

Chapter Knowledge

1. I know that about half my grade is based on previous math knowledge and ability to learn. ____
2. I know if I was placed into my first math course without a placement test a math assessment is needed to determine my best math course. ______
3. I know that up to 25% of my grade is based on instruction and tutoring. _____
4. I know that between 25% and 41% of my grade is based on affective charateriateis such as study skills, anxiety, motivation and self-concept. _____
5. I know that controlling my math success is primarily based on improving strategies that affect my study skills, anxiety, motivation and self-concept (affective characteristics). ____
6. I know that if I have test anxiety to talk to the advisor, counselor or instructor and read Chapter 3 (Managing and Reducing Math Anxiety and Test Anxiety). ____
7. I understand that using my best learning modality can improve my memory. _____
8. I understand that by combining two of my best learning modalities it can greatly enhance my memory. ______
9. I understand that common memory techniques include maintaining a good math/study attitude, becoming a selective learner, becoming on organizer, using visual imagery, making associations, using mnemonics devices and acronyms. _______
10. I used the metacognition model of planning, monitoring and evaluating to improve my math success. _____
11. I completed My Math Success Plan. ____
12. Base on My Math Success Plan I know which math success strategies to improve. _____

Behavioral Activities

1. I went to the math lab, tutor center, Learning Resource Center, or websites to find resources to improve my math background on this date / / . _____
2. I reviewed the Math Study Skills Evaluation - C results and based on my scores decided to first learn the information in Chapter(s) _______ and ______.
3. My best learning modality is ______ and I discussed with my tutor how to best to tutor me based on my style on this date / / .
4. I talked to my classmates about their learning styles to discover their strategies. _____
5. I shared My Math Success Plan with my counselor, teacher, ROTC advisor, or parent and discussed how to improve math success. ____

Name: ______________________________ Date: ______________

Assignment for Chapter 2

On a separate sheet of paper, answer the following questions:

1. How does math knowledge affect your math grades?

2. How does quality of instruction affect your math grades?

3. How does affective characteristics affect your grades?

4. List and describe the three different types of learning modality styles?

5. List you best modality learning style and describe five different strategies to from that style to improve learning.

6. How does your learning style match up with your current professor's? If there is a mismatch what can you do?

7. List two general memory techniques and describe how you use them to remember math formulas or concepts.

8. List and describe how you would use the three steps in metacognition to solve math problems.

9. Describe strengths and areas to improve on in your "My Math Success Plan".

10. List five strategies that you will implement this week to improve on from your "My Math Success Plan".

Managing Math and Test Anxiety

3

In Chapter 3 You will learn these concepts:

- ✓ Understanding the causes of math anxiety
- ✓ The different types of test anxiety
- ✓ The effects of test anxiety on learning
- ✓ How to reduce test anxiety
- ✓ How to replace negative self-talk with positive self-talk

Understanding Math Anxiety

Math anxiety is a relatively new concept in education. During the 1970s, certain educators began using the terms "mathophobia" and "mathemaphobia" as a possible cause for a child's unwillingness to learn math. Modern psychologists define math anxiety as an extreme emotional and/or physical reaction to a negative attitude toward math.

Math anxiety affects students in many different ways. It sometimes manifests as tension, which interferes with the manipulation of numbers and the solving of math problems during tests (Richardson and Suinn, 1973). It also includes the panic, helplessness, paralysis and mental disorganization that occurs in some students when they are solving math problems. This discomfort varies in intensity and results from previous experiences in past learning situations (Tobias, 1978). There is a strong relationship between low math confidence and high math test anxiety.

Math anxiety is common among college students. Counselors at a major university recently reported that one-third of students enrolled in behavior therapy programs—usually offered through counseling centers—have problems with math anxiety. It has been shown that math anxiety exists among many students who do not suffer from other tensions. It frequently occurs in students with a poor high school math background. Approximately half of students enrolled in developmental math courses suffer from some form of math anxiety. Though students in high-level courses occasionally show symptoms as well, symptoms are usually worse for developmental students. Today, math anxiety is accepted as one of the major problems students have in completing their math courses. It is real, but it *can* be overcome.

Math anxiety is divided into three separate anxieties: Math Test Anxiety, Numerical Anxiety and Abstraction Anxiety. Math Test Anxiety involves the anticipation, completion and feedback of math tests.

Numerical Anxiety refers to everyday situations that involve numbers and arithmetic calculations.

Numerical anxiety affects students who are trying to figure out the proper amount for a tip, thinking about mathematics, doing math homework or listening to a math lecture.

Abstraction Anxiety involves working with variables and mathematical concepts used to solve equations. Some students experience all three types of math anxiety; others deal with only one.

Most of the students I have worked with deal with Math Test Anxiety and Abstraction Anxiety. These students do not have any anxiety with numbers, but often struggle with algebra and other symbol-heavy math. Unfortunately, these symbols are extremely common in college math courses.

Causes of Math Anxiety

Since math anxiety is a learned condition, its causes are unique to each student. Nevertheless, math anxiety is almost always rooted in an individual's past experiences. Bad experiences in elementary school are one of the most common sources for students' math anxiety: coming in last in math races at the blackboard, watching a classmate finish a problem twice as fast as they do, or teachers saying, "That's okay. You just aren't good in math; you are better in English." When asked, many students with math anxiety indicate that they were made fun of when trying to solve math problems at the chalkboard. This is by far the most common story I hear when working with students. Heckled students have a hard time letting go of painful memories. Harsh words and experiences remain with these students; when they walk into a classroom or open a math book, these bad experiences often play back in their minds.

A 56 year old once indicated to me that he had a great deal of fear that his instructor might call him to the board. Even if he knew how to do a problem, he was terrified at the thought of having to display that knowledge to his peers.

Example: Over the years, math anxiety is reinforced and even increases in magnitude. In fact, many math anxious students – now 30, 40 and 50 years old – still have extreme fear about working math problems on the board. These bad memories linger well into adulthood.

Being embarrassed by family members also causes math anxiety. According to a recent study, many students claim that their parents routinely try to help them with math and this sometimes leads to serious trauma. These students say that tutoring from their guardians, mainly their fathers, often results in scolding when they are not able to complete problems. One student reported that his father hit him every time he got a multiplication problem wrong.

Brothers and sisters also tease one another about being dumb in math. This is particularly true of boys telling girls that they cannot do math. When people hear these statements enough times, they often start to believe them and associate these bad feelings with math. For some students, just hearing the word "math" triggers a response of anxiety as they consciously or unconsciously recall bad feelings or memories.

A good example of this is a student who I worked with who had completed her BS degree 15 years ago at a college that did not require much math. She was returning to college to be an elementary school teacher, which required her to take math and a placement test. As soon as I mentioned that she had to take math, she said, "I can't do math, and I will have to wait a few days to get psychologically ready to take the math placement test." She indicated her old anxiety feelings rushed through her, and she almost had an anxiety attack. This is an extreme but true example of math anxiety. In most cases math anxiety is not this bad, but remains disruptive enough to cause learning and testing problems.

Those who don't have math anxiety still have to understand it. This way they can help their classmates who need support. Also, if you do not have math anxiety now, you may develop it in the future. One way to overcome math anxiety is to find out when it first occurred and how it is still affecting you. A math autobiography is an excellent way to review previous math experiences and to learn how to overcome math anxiety. To find out more about writing a math autobiography, complete the first two activities in this chapter.

How Math Anxiety Affects Learning

Anxiety causes learning problems in several ways. Let's first start by looking at how it affects your homework. Students with high math anxiety have difficulty starting or completing assignments. Doing homework reminds some students of their learning problems in math. More specifically, it reminds them of previous math failures, which causes further anxiety. This anxiety sometimes leads to total avoidance of homework or "approach-avoidance" behavior; students start their homework, then they quit, return later, then quit again. Total homework avoidance is called procrastination. The thought of doing homework causes these students anxiety, which then causes them to put it off. Procrastination makes them feel better for a short time – until test day.

Math anxiety also affects your classroom participation and learning. Usually students with math anxiety are afraid to speak out in class and ask questions. They are also afraid of asking a question that others, including the teacher, might consider dumb. They sit in class fearful of being asked a question, looking like they understand the lecture so they will not be called on. They also take a lot of notes, even though they don't understand them, to give the illusion of comprehension. If you are one of these students, these are hard habits to break. Still, there are numerous ways to overcome them. Here are just a few:

1. *Make an appointment to talk to your math teacher.* Math instructors want to talk to you. When I do my consulting around the country, one of the major complaints I get from math instructors is that students don't come and see them. Make an appointment to see your math instructor before the first major test to discuss your math history and to ask for suggestions.
2. *Before class, ask the teacher to work one homework problem.* Write the problem on the board before the instructor arrives. This is less stressful because you are not asking the question in front of the whole class. Choose an easy question, if possible.
3. *Prepare one question from your homework and ask it within the first 15 minutes of class.* Instructors are more likely to answer questions in the first part of class when they have time instead of the end of class when time is running out.
4. *Ask a question about a problem to which you already know the answer.* That way, if the instructor asks you a question about the problem, you will know the answer. This is good practice for asking questions about things you don't understand.
5. *Use email to send questions to your instructor.* This way you can still get the answer with very little anxiety. Do not, however, overuse this option. If you send too many emails, your instructor may begin to ignore you.
6. *Set up an appointment with your disability counselor/advisor to discuss anxiety issues.* Review your test scores from your psycho-educational report, IEP or Section 504 plan. If necessary, ask for a referral to obtain help in reducing anxiety. Do not be embarrassed. Always remember to get the help you need.
7. *Talk to your parents about your anxiety.* Discuss your anxiety with your parents, as well as how you can get help to reduce your anxiety. They may suggest seeing your doctor or counselor.

By completing these steps, you are setting up a new history of positive expriences. Not only will these tips help you become more comfortable in a math classroom, but they also allow you to change your overall attitude toward math. It is important to remember that math anxiety *can* be overcome with your effort. You don't have to live in the past with your math fears. Today is a new day and you can change how math affects you.

Activity 3.1 Revisiting Past Math Experiences

If you have math anxiety, try to remember the first time you had uneasy feelings about math. This does not include anxiety when taking a math test, which will be discussed later in this chapter. To help you remember this experience, check the appropriate response and answer the following questions:

1. Was your first negative math experience in:

   Elementary School _____

   Middle School _____

   High School _____

   Never _____

2. Can you recall the incident(s)?

   Yes _____

   No _____

3. Was it after....

   Being called on in class and getting the answer wrong? _____

   Getting a poor grade on a homework assignment? _____

   A parent saying he/she cannot do math and neither can you? _____

   Another adult telling you that cannot do math? _____

   A fellow student telling you that you are not good at math? _____

   A teacher telling you that you are not good at math. _____

4. If your first negative math experience is not listed above, then write it down here:

5. If you cannot remember a specific incident, then when was the last time you told yourself that you could not learn math?

   Today _____

   Yesterday _____

   Last Month _____

   Last Year _____

   Never _____

Understanding Test Anxiety

Test anxiety has existed for as long as tests have been issued to evaluate student performance. Because it is so common, and because it has survived the test of time, test anxiety has been carefully studied over the last fifty years. Pioneering studies indicate that test anxiety generally leads to low test scores.

At the University of South Florida (Tampa), Dr. Charles Spielberger investigated the relationship between test anxiety and intellectual ability. The study results suggested that anxiety coupled with high ability can improve academic performance; but anxiety coupled with low or average ability often interferes with academic performance. That is:

Anxiety + High Ability = Improvement
Anxiety + Low or Average Ability = No Improvement

Test anxiety is a learned response brought on by environtmental conditioning. A person is not born with it. It is a special kind of general stress, or a "strained exertion," which sometimes leads to physical and psychological problems. The good news? Because test anxiety is a *learned* response, it can eventually be *unlearned* through hard work and discipline. Read on through this chapter to find out more.

Definition of Test Anxiety

According to the Diagnostic and Statistics Manual of Mental Disorders, test anxiety is characterized by an "extreme fear of poor performance on tests and examinations." The DSM goes on to describe a statistical correlation between test-anxiety and various character attributes, including but not limited to low self-esteem, dependency and passivity.

One of my students once compared test anxiety to "being in a burning house with no way out." Another described it as "a sick feeling I get on test days that makes me feel like a child and makes me forget everything."

No matter how you define it, test anxiety is real, and it affects millions of students.

The Causes of Test Anxiety

The causes of test anxiety are different for each student. The initial incident could possibly have occurred in middle or high school. For many students, however, test anxiety first occurs in college when passing tests is the only way to pass a course. In most college courses, homework and extra credit do not count toward your grade. Now students must have a passing average, and, in some cases,pass an intimidating departmental final exam.

Additional pressure also exists because not passing algebra means you won't graduate, and you might not get the job you want.

PTSD is another cause of test-anxiety. As described previously, students with PTSD often fear stress itself, which in turn causes a vicious loop of self-doubt and panic. Many of these students are also dealing with Traumatic Brain Injuries, which often go hand-in-hand with memory problems. Students with TBIs are often caught in a viscious loop. First, they fear that they will forget important information on test days. Then, on test day, the stress from this fear causes them to struggle, which only reinforces their initial anxiety.

Since we have already explored your experiences with general math anxiety, let's look at some of the direct causes of your math test anxiety. If you do have test anxiety, what is the main cause? If you don't know, then review the eight causes of test anxiety on the next page. Does one of these reasons fit you? If you don't have test anxiety, think of scenarios that might bring it on in the future.

If you have math test anxiety, following the suggestions in this chapter and the rest of the book can greatly reduce it. The first step was already taken by understanding how you developed test anxiety. The second step is writing the reasons.

For conveinence, here is a review of the 10 basic causes of test anxiety.

1. *Test anxiety is* a learned behavior resulting from the expectations of parents, teachers or other significant people in the student's life.
2. *Test anxiety is* sometimes caused by the association between grades and a student's personal worth.
3. *Test anxiety* develops from the fear of alienating parents, family or friends due to poor grades.
4. *Test anxiety* stems from a feeling of lack of control and an inability to change one's life situation.
5. *Test anxiety* is sometimes caused by a student being embarrassed by the teacher or other students when trying to do math problems.
6. *Test anxiety* often occurs during timed tests due to the fear of not finishing the test. This is true even when a student is capable of doing all of the problems.
7. *Test anxiety* is sometimes caused by being put in math courses that are above the student's level of competence.
8. *Students leaving* the room before the test time is up often triggers anxiety among the remaining students.
9. *PTSD* often triggers test-anxiety.
10. *TBIs* often trigger test-anxiety.

The Different Types of Test Anxiety

The two basic types of test anxiety are *emotional* ("somatic") and *worry* ("cognitive"). Students with high test anxiety have both emotional and worry anxiety.

Signs of emotional anxiety are upset stomach, nausea, sweaty palms, pain in the neck, stiff shoulders, high blood pressure, rapid shallow breathing, rapid heartbeat or general feelings of nervousness. As anxiety increases, these feelings intensify.

Even though these feelings are caused by anxiety, the physical response is real. These feelings and physical inconveniences affect your concentration and your testing speed, and sometimes cause you to "draw a blank."

Worry anxiety causes students to think about failing a test. This negative "self-talk" causes students to focus on their anxiety instead of recalling math concepts by telling themselves that they will fail the test. This talk is perhaps the worst enemy of a math student. Not only does it reinforce bad behavior, it also prohibits a student from establishing positive energy. Negative self-statements vary from student to student.

Students with worry anxiety have told me many things after not being able to solve problems:

- "What is wrong with me?"
- "I did these types of problems before."
- "I just cannot get math!"
- "Math is awful!"
- "I hate math!"
- "I don't need it for my career so why am I taking the course."
- "I am going to fail this test and this course."
- "I might as well drop this course and quit college."
- "I am dumb."

These types of statements cause more worry anxiety and divert your attention away from the test, leaving you less time to complete problems. Later on in the chapter we discuss how to get rid of these types of statements.

Students may have different levels of emotional anxiety and worry anxiety. The treatment depends on the levels of your type of test anxiety. All students have a total test anxiety level, a emotional anxiety level and a worry level. On a scale of 1 to 100, determine your self-reported total anxiety, worry anxiety and emotional anxiety. A level of 1 means no anxiety, a level of 50 means average anxiety and a level of 100 means the highest anxiety (meaning you cannot even walk in the room during test day without getting sick). Students usually have different anxiety levels in all three categories. Below, fill in the numbers that represent your levels.

Emotional anxiety _____

Worry anxiety _______

Total anxiety _____

Low levels of anxiety are between 1 and 25. Moderate low levels of anxiety are between 26 and 50. Moderately high levels of anxiety are between 51 and 75. High levels of anxiety fall between 76 and 99. Based on this information you can change your score.

What is your highest area of anxiety? Students who have high test anxiety have both emotional and worry anxiety.

The Effects of Anxiety on Testing and Learning

The effects of anxiety on student learning and testing can best be explained by looking at the Stages of Memory process. The Stages of Memory are Sensory Input, Sensory Register, Short-term Memory, Working Memory, Abstract Reasoning, Long-term Memory and Memory Output.

When anxious students are reading a math textbook or learning mathematics in the classroom, their anxiety affects their Sensory Register and Short-term Memory. Anxiety interferes with how fast people process information and can decrease the amount of information they can hold for a short period of time. This means that less information enters Working Memory, which is where mathematics information is combined and learned.

> Example: Anxiety manifests itself differently in every student. The effects of test anxiety range from a "mental block" on a test to avoidance of homework. These symptoms also range in severity.

Also, anxiety affects Working Memory by decreasing the amount of information that can be processed at the same time. This leads to poor Memory Output such as not understanding what you read or not answering a question during class.

The most recent research indicates that test anxiety strongly affects Working Memory. Authors Ashcraft and Kirk (2001) indicate that math anxiety temporarily disrupts mental processing in Working Memory that causes poorer math achievement. Math anxiety uses up Working Memory resources that make it harder to learn math. During tests, math test anxiety decreases the amount of Working Memory space, which means less information can be received and used from Long-term Memory and Abstract Reasoning. This is like using a calculator in which half of the keys are not functioning. The result is a slow down in performance and a decrease in accuracy, resulting in poorer grades.

Test anxiety often makes it more difficult for math students to recall formulas and concepts on exams. This causes stress, which only strengthens anxiety.

One of the most common side effects of test anxiety is getting the test and immediately forgetting information that you know. Some students describe this event as having a "mental block," "going blank," or indicating that the test looks like Greek. After five or ten minutes into the test, some of these students can refocus and start working the problems. They have, however, lost valuable time.

For other students, anxiety persists throughout the test and they cannot recall the needed information. It is only after they walk out the door that they can remember how to work the problems. When this happens, they get mad at themselves, which increases their fears that the same thing will happen on the next test.

Sometimes students with math test anxiety do not "go blank," but it takes longer to recall formulas and concepts and to work problems. The result is frustration and loss of time, leading to more anxiety.

Since, in most cases, math tests are speed tests (those in which you have a certain amount of time to complete the test), you may not have enough time to work all the problems or to check the answers if you have mentally slowed down. The result is a lower test score because even though you knew the material, you did not complete all of the questions before the test time ran out.

Not using all of the time allotted for the test is another problem caused by test anxiety. This is particularly frustrating, because this behavior is entirely avoidable. Most students know that they should use all of their test time to check their answers. In fact, math is one of the few subjects in which you can check test problems to find out if you have the problems correct. However, most students do not use all of their given test time, and this results in lower test scores.

Why? Students with high test anxiety do not want to stay in the classroom. This is especially true of students whose test anxiety increases as a test progresses. Test anxiety gets so bad in these situations that some students are willing to get a bad grade just to leave the classroom a little bit early.

Even more common among anxious students is the fear of what the instructor and other students will think about them for being the last one to hand in a test. These students refuse to be in the last group to finish, because they feel the instructor or other students will think they are dumb.

This is middle-school thinking, but the feelings are still real – no matter the age of the student. These students do not realize that some students who turn in their tests first fail, while many students who turn in their tests last make "A's" and "B's." Assumptions in this regard are extremely detrimental toward making good grades.

When taking a test, you need to focus on

yourself and yourself only. Do not worry about what your classmates are doing. Instead, keep your eyes on your own test and imagine the relief you will feel when you hand in your test confident that you did not make any careless errors.

Another effect of test anxiety relates to completing homework assignments. Students who have high test anxiety often have difficulty starting or completing homework. The material on the homework causes them to think about previous tests or begin worrying about the next test and whether they are going to stress out in front of their classmates.

Some students begin their homework and work some problems successfully. They then get stuck on a problem that causes them anxiety, so they take a break. During their break the anxiety disappears until they start doing their homework again. Doing their homework causes more anxiety, which leads to another break. The breaks become more frequent. Finally, the student takes one long break and does not do the homework. Quitting, to them, means no more anxiety until the next homework assignment.

The effects of math test anxiety can be different for each student. Students can have several of the mentioned characteristics that can interfere with math learning and test taking. Remember, many popular conceptions about test anxiety are entirely inaccurate. Review the 12 Myths About Test Anxiety chart, printed to the right, to see which ones you believe.

It is important to stick to the information in this book and not fall victim to advice from well-meaning but uninformed friends. Test anxiety is a very real problem and cannot be cured by simply disregarding its existence.

Now that we understand how math anxiety and math test anxiety affect learning and testing, it is time to learn how to reduce your test anxiety. The first step to doing this is to remain positve.

Tell yourself, "I can reduce my math test anxiety. It may take some time, but I *can* reduce my math test anxiety."

12 Myths About Test Anxiety

1. Students are born with test anxiety.
2. Test anxiety is a mental illness.
3. Test anxiety cannot be reduced.
4. Any level of test anxiety is bad.
5. All students who are not prepared will have test anxiety.
6. Students with test anxiety cannot learn math.
7. Students who are well prepared will not have test anxiety.
8. Very intelligent students and students taking high-level courses, such as calculus, do not have test anxiety.
9. Attending class and doing my homework should reduce all of my test anxiety.
10. Being told to relax during a test will make you relaxed.
11. Doing nothing about test anxiety will make it go away.
12. Reducing test anxiety will guarantee better grades.

Quick Section Review:

- *Test anxiety* manifests itself in many different ways.
- *There are two* types of test anxiety: emotional and worry.
- *Test Anxiety* is known to affect working memory. This diverts mental resources away from math and causes numerous mental blocks.
- *Anxious students* should make sure to use all of their alloted test time.
- *Test Anxiety* also affects homework sessions when the math on the assignment reminds them of bad test experiences.

Activity 3.2 Revisiting Past Test Experiences

If you have test anxiety, try to remember the first time you had uneasy feelings about taking a math test. To help you remember this experience, check the appropriate response and answer the questions:

Was your first negative test experience in:

Elementary School _____

Middle School _____

High School _____

1. Can you recall the incident(s)?

   Yes _____

   No _____

2. Was it after....

   Your first alegbra test? _____

   Your first math test after being out of school for a long time? _____

   After you decided to get serious about college? _____

   When a professor told you that you needed to pass a test to pass a course? _____

   When you needed to pass a test to maintain your financial aid? _____

   After a parent asked you why you failed a previous test? _____

   Your children asking you why you failed your last test? _____

   If your first negative test experience is not listed above, then write it down here:

If you cannot remember a specific incident when you had test anxiety, do you expect to have any major test anxiety on your next math test? If so, write down your situation below:

How to Reduce Math and Test Anxiety

To reduce math and test anxiety, you need to understand both the relaxation response and how negative self-talk undermines your abilities. These anxiety reduction techniques and positive self-statements can be used before or during your homework, class or test. These techniques need to be practiced several times before they become effective. If you have extremely high test anxiety, you need to practice long-term relaxation techniques. This mainly involves learning how to turn negative self-talk statements into positive self-talk statements.

Short-Term Relaxation Techniques

The relaxation response is any technique or procedure that helps you to become relaxed and takes the place of an anxiety response. There are both short-term and long-term relaxation response techniques, which help control emotional math test anxiety.

These techniques also help reduce worry anxiety. Among others, effective short-term techniques include the Tensing and Differential Relaxation Method and the Palming Method. The next four subsections describe these methods in further detail. Try them all to figure out which works best for you.

Tensing/Relaxing Method

The Tensing and Relaxation Method helps you relax by tensing and relaxing your muscles all at once. Follow these procedures while sitting at your desk before taking a test:

1. *Put your* feet flat on the floor.
2. *Grab underneath* your chair with your hands.
3. *Push down* with your feet and pull up on your chair at the same time for about five seconds.
4. *Relax* for five to ten seconds.
5. *Repeat the procedure* two to three times.
6. *Relax all your muscles* except the ones that are actually used to take the test.

Deep Breathing Technique

Deep Breathing is a technique that can reduce your test anxiety. Follow these steps to Deep Breathing:

1. *Inhale slowly* and deeply through your nose by filling up the bottom of your lungs first.
2. *Stop* for a few seconds and hold your breath.
3. *Exhale slowly* through your mouth pretending like you are whistling out the air. Be sure to exhale fully and let your whole body relax. Wait a few seconds and then start to inhale as stated in number 1.

After practicing deep breathing it should become a very natural process. You may want to repeat the steps five times and then take a break. You may then want to do one more set.

If you start to get light-headed while practicing, stop for a few minutes. You were probably breathing in and out too fast. Practicing this breathing technique will make it easier to use before or during the test if needed. This technique will help some students to relax, while other students may prefer other short-term techniques.

Visualization

Have you ever daydreamed in class or during a workshop? I have asked this question to thousands of students and the answer is always yes. Then I ask them what happens when they daydream. The answer is that their minds leave the room and the body thinks it is where the mind has gone. Then I ask what would happen if you daydreamed about some relaxing place? Your mind would tell your body to relax. This process is called visualization, or using visual images in your mind to reduce test anxiety.

This technique has been used for hundreds of years. Athletes also use this technique to reduce their anxiety and to improve their game performance. You can imagine anything or anyplace in order to reduce your test anxiety. Follow these steps to practice visualization:

1. *Find* a comfortable place to sit.
2. *Close your eyes* and *think* about a relaxing place, real or imaginary.
3. *Imagine yourself* in that place, making sure to *see* it through your own eyes, not seeing yourself from afar.
4. What type of sounds do you *hear*? What do you *smell*? Are you feeling the sand between your toes or maybe the cool air? *Make the scene* as real as possible.
5. *Visualize that scene* for one to two minutes.
6. *Open your eyes* and continue to feel relaxed.

> Example: You need to develop two scenes and practice them every day until you can relax yourself in a few minutes. Then you will be ready to use them on the test.

Palming Method

The palming method is a visualization procedure used to reduce test anxiety. While you are at your desk before or during a test, follow these procedures:

1. *Close* and *cover your eyes* using the the palms of your hands.
2. *Prevent your hands* from touching your eyes by resting the lower parts of your palms on your cheeks and placing your fingers on your forehead. Do not rub or touch your eyeballs.
3. *Think* of some real or imaginary scenes that are relaxing to you.
4. *Visualize* this scene. Picture it as if you were actually there. Do this for one to two minutes.

Practice visualizing this scene several days before taking a test and the effectiveness of this relaxation procedure will improve.

Long-Term Relaxation Techniques

The Cue-Controlled Relaxation Response Technique is the best long-term relaxation technique. Cue-Controlled Relaxation works by inducing your own relaxation based on repeating certain cue words to yourself. In essence, you are taught to relax and then silently repeat cue words, such as "I am relaxed." After enough practice, you can relax during math tests. To practice these techniques, talk to your advisor or counselor for resources.

Managing Self-Talk

Imagine two students taking their first math test during the semester, and half way through the exam they start missing several problems. One student starts saying to himself that he is going to fail the test and might as well turn in his paper and quit. The other student says: "I might miss these problems but that is not a reason to give up and leave. I will just try as hard as I can on the other problems and I can pass."

One student feels a lot of anxiety and the other student remains calm and develops a plan. In both cases, the situation is the same, but the feelings in response to the test situation are extremely different due to their internal dialogue or self-talk. Which student are you?

Cognitive psychologists claim that what we say to ourselves in a response to an event mainly determines our mood or feeling about that event. Sometimes we say these statements to ourselves so quickly and automatically that we don't even notice. We then believe the situation is causing the feeling, when it is actually our interactions or thoughts about the event that are controlling our emotions.

This sequence is represented by the following timeline:

External Events (Math Test)

↓

Interpretation of Events with Self-Talk

↓

Feelings, Emotions and Reactions

Based on this psychological theory, in most cases you are responsible for what you feel. You have a choice to have positive or negative self-talk while doing your math homework or during a math test. This can determine how you feel during the test and control some of your anxiety. Some students see a math test as an opportunity to show their knowledge of the subject and are excited about that, while others see the same test as a potential failure that will lead to anger and dissatisfaction. The realization that you are responsible for your own feelings is very powerful once you fully accept it. Some points about self-talk include:

- Self-Talk is like a telegraphic message where one or two words can bring up many different thoughts and feelings.
- Self-Talk is like a scratched CD. It plays the same words over and over again.
- Self–Talk happens automatically. You don't even think about what you are saying to yourself.
- Negative self-talk during anxious situations is usually illogical, though it might seem rational at the time.
- Positive self-talk can increase appropriate behavior. Telling yourself that you can do your homework can lead to actually doing your homework.
- Negative self-talk causes avoidance and procrastination. Students who tell themselves they are not good in math don't want to do their homework and avoid it whenever possible.

Negative Self-Talk

Students who have high test anxiety usually use negative self-talk. This negative self-talk can increase the student's test anxiety and may cause them to fail. It is totally possible for students with negative self-talk to change this bad habit.

To change the negative self-talk habit, it is beneficial to understand the different types of negative self-talk. If you have negative self-talk, then review the different types to see which one matches you best. You may be one type or a combination of different types.

The Worrier— Worriers always look for the worst-case situation. They are scared of failure. When they feel a little bit of anxiety, they blow it out of proportion and give up, believing there is nothing that can help them pass. In some extreme cases, they drop out of college believing they will never be able to pass a math course. The worrier's favorite question is, "What if . . . ?" For example, "What if I fail this test and then this class? I will not graduate and be in this dead-end job forever or all my friends will make fun of me." The Worrier can eventually give in to fear and sabotage his/herself in self-defense from disappointment.

The Victim— "Victims" want to feel helpless and hopeless. They create anxiety by telling themselves that no matter what they do, they will not be successful in math. Victims believe there is something wrong with them that is not curable. They do not blame other factors that they can change, such as decreasing anxiety and improving their math study skills. They want to doom themselves and get into a learned helplessness mode that eventually stops them from even trying. They grow accustom to failure. The Victim's favorite statement is, "I can't. I will never be able to no matter how hard I try . . ."

The Critic— Critics like to put themselves down. They look for internal flaws, and if

Negative self-talk causes students to behave in a counterproductive manner. A proper self-esteem is required for success in math classes.

they don't have any, they create some. They put themselves down when they notice other students not showing any anxiousness. They ignore their success, and instead of being proud of their accomplishments, they believe them to be a fluke. The Critic's favorite statement is, "You are too stupid to learn math!" If Critics do pass a test, they say afterward, "You could have made a better grade. Look at all those careless errors." The Critic's goal is to promote low self-esteem so that he or she will stop learning math. This behavior is disasterous come grade time. More often than not, it leads to failure.

The Perfectionist

Perfectionists are closely related to Critics, but instead of putting themselves down for missing problems, they use the experience to push themselves toward success.

Perfectionists cannot stand mistakes or poor grades. They drive themselves to exhaustion making sure they understand every single concept. Self-worth means nothing to them because they will never be happy with anything less than perfection. This eventually leads to failure when they pass their "breaking point." Even a grade such as a 92 is not good enough. In some cases, it drives them toward causing harm to themselves.

Do any of these sound familiar?

After reading about these different types of personalities, which one can you most relate to? The person inside of you may be a combination of two different types, but there usually is one personality that dominates your perception of math. Once again, do not beat yourself up if you fall into one of these categories. In fact, just about everybody exhibits signs of at least one category. With this in mind, the next step is to learn how to cope with these problems, as well as how to begin using your personality type to become successful.

Positive Self-Talk

You can counter and control negative self-talk in several ways. Negative self-talk is easily replaced with positive self-talk. You can also develop thought-stopping techniques that reduce negative self-talk and engender positive self-talk.

Try all of the following ways to discover which one, or combination of two techniques, work best for you. The first technique is to develop positive self-talk statements. The following list of suggestions will help you develop a series of personal statements, which you should memorize as soon as possible.

1. *Use the first person tense.* For example, "I can control my anxiety and pass this test."
2. *Avoid using negatives in the statement.* For example, don't say, "I will not get nervous during this test." Instead say, "I will calm myself down during the test.
3. *Make the statements positive and realistic.* For example, you can say, "I will be successful on this test", instead of saying, "I will make a 100 on this test."

The Worrier, who asks, "What if I fail the test?" can counter with, "If I fail the test I will just do better on the next one."

The Victim, who believes things are hopeless says: "No way will I ever pass math." They can counter with: "I used a different way to study and take the test so I can pass the course this time."

The Critic, who puts themselves down by saying: "I cannot reduce my test anxiety and will fail." They can replace this by saying: "I know I have test anxiety but I have learned to control it."

The Perfectionist, who says, "I must make an A or I am a failure," can instead say, "I don't need an A to be successful. I just need to pass this course so I can take the next one and graduate."

These are examples of positive self-talk that can control anxiety. Additional positive self-talk statements are listed below.

- "I failed the course last semester, but I can now use my math study skills to pass this course."
- "I went blank on the last test, but I now know how to reduce my test anxiety."
- "I know that my poor math skills are due to poor study skills, not my own ability, and since I am working on my study skills, my math skills will improve."
- "I know that, with hard work, I will pass math."
- "I prepared for this test and will do

the best I can. I will reduce my test anxiety and use the best test-taking procedures. I expect some problems will be difficult, but I will not get discouraged."

Create your own positive self-talk statements in Activity 3.3, which is found at the end of this section.

Thought-Stopping Technique

Many students have difficulty controlling their negative self-talk. These students have tried to tell themselves to eliminate the negative self-talk, but no matter what they try it persists. These students need a thought-stopping technique to break this bad habit.

Thought-stopping entails focusing on unwanted negative self-talk, and then suddenly stopping those thoughts with some type of internal or external action. Actions such as yelling to yourself, "Stop that," or making a loud noise, such as slapping a desk, effectively interrupt negative self-talk.

Obviously, your location plays a role in what you should and should not do. In a homework situation, you may want to slap the desk, but obviously doing this in a crowded classroom is inappropriate.

To stop your negative thoughts in a crowded classroom, while listening to a lecture, or while during a test, silently shout to yourself, "Stop thinking about that." After your silent shout, either relax yourself or repeat one of the positive self-talk statements that you have made up. You may have to shout to yourself several times to control your negative self-talk. After every shout, use a different relaxation technique, such as positive visual scenes or positive statements that will help control your anxiety.

The way "stop-thinking" works is by interrupting the worry response before it creates the type of anxiety that gets out of control. During the interruptions, you gain control and replace negative self-talk with positive responses. Students with high worry anxiety need to practice these techniques at least one week before a test several times a day. Then, they need to keep practicing once a day until the negative self-talk completely disappears.

If needed, obtain additional help from your counselor or college psychologist to help you stop negative self-talk. Doing so prepares you to deal with inevitable adversity and sets you on the right path toward a positive math attitude.

Power of Positive Thinking

If you remain skeptical about the power of positive thinking, here is one more fact to persuade you. According to a 1985 study, conducted by psychologists Michael Scheier and Charles Carver, optimists are much more likely to succeed in life, largely because they are more capable of coping with adversity.

When they are presented with challenges, optimists manage to navigate through their negative feelings and ultimately persevere. The benefits of being an optimist don't stop there. They also live longer and sustain better overall mental and physical health than do pessimists.

Mindfulness

Another great technique to reduce anxiety involves the "Mindfulness" techniques mentioned in Chapter One. If you remember, mindfulness is the ability to interact with the present moment with full attention, without allowing thoughts, internal dialogue, and accompanying emotions to interfere with how we interact in the present.

In many ways, mindfulness is just another way to ensure that you maintain a happy and healthy relationship with your high school/ college experience. Oddly, graduates often look back fondly at stressful tests. Eventually these tests become war stories: the sort a person often embellishes to either get a laugh or paint a vivid picture of a difficult course. By staying in the moment, you can appreciate stressful times as being part of the overall excitement of being in college.

Activity 3.3 Creating Positive Self-Talk Statements

As you've just read, in order to succeed in a math class, you must replace negative self-talk statements with positive self-talk statements. Now that you've read the instructions in this chapter, you can go ahead and develop your positive self-talk statements in the space below. Develop four positive self-talk statements that are not in this book, making sure to use the word "I "in each statement.

Statement 1:

Statement 2:

Statement 3:

Statement 4:

Chapter Three: Checklist for Knowledge and Behaviors

Put a check after understanding the knowledge statements and fill in the behavioral activities.

Chapter Knowledge

1. I understand that math anxiety is usually caused by past experiences. _____
2. I understand that math anxiety affect learning in class and doing homework._____
3. I understand that math anxiety may cause me to procrastinate in doing homework and preparing for tests. ____
4. I know that test anxiety is a learned behaviors developed by having emotional and /or worry response during previous tests. _____
5. I know I can unlearn general and test anxiety behaviors. _____
6. I know that test anxiety often occurs during timed test due to a fear of not completing._____
7. I know that some test anxiety can improve performance however to much test anxiety can decrease performance. ____
8. I understand the two types of text anxiety (emotional, worry) and how they can affect me in different ways. _____
9. I know that test anxiety reduces my ability to complete test because it blocks memory and causes an urgency to leave the room before checking all my answers. _____
10. I know that I must learn relaxation techniques to help reduce my anxiety. _____
11. I know that reducing my test anxiety does not guarantee success on tests; fist I must know the appropriate material to recall during the test _____
12. I know can improve my math study skills and increase my study time to build positive experiences that will support me during the test. _______

Behavioral Activities

1. I will set up an appointment with my counselor, advisor, or math teacher to discuss test anxiety. ____
2. I have selected my two best short-term relaxation techniques and will practice them three times a week. ___
3. I will manage my self-talk about math by reducing my negative statements and changing them to positive statements. _____

Name: ________________________________ Date: ________________

Assignment for Chapter 3

1. Create your personal definition of math anxiety.

2. List and explain three causes of math anxiety. Which cause would you most likely have?

3. Describe two different ways math anxiety can affect learning

4. List and describe three ways to help overcome math anxiety.

5. Describe how you know if you have math anxiety or math text anxiety.

6. List and describe five causes of test anxiety.

7. List and explain the two different types of test anxiety.

8. Describe your best short-term relaxation technique.

9. List three positive statements you can tell yourself during the test.

10. Even if you may not have test anxiety describe your plan to reduce anxiety for the next test.

How to Improve Listening and Note-Taking Skills

4

In Chapter 4 You will learn these concepts:

✓ How to listen effectively and identify important concepts

✓ How to design a classroom note-taking system

✓ How to design an online and Emporium note-taking system

✓ How creating a math glossary improves learning

✓ The importance of emailing instructors

How toBecome an Effective Listener

Listening and note-taking skills in a math class are very important, since most students do not read the math text or have difficulty understanding it. In most of your other classes, if you do not understand the lecture, you can read the book and get almost all the information. In the math class, however, the instructor can usually explain the textbook better than students can read and understand it. Students who do not have good listening skills or note-taking skills are at a disadvantage in learning math. Most math understanding takes place in the classroom. Students must learn how to take advantage of learning in the classroom by becoming effective listeners, calculator users and note-takers.

You can become an effective listener using a set of skills, which you can learn and practice. To become an effective listener, you must prepare both physically and mentally.

It is also important to point out, students are now receiving their lectures online. For example, you may be in an Emporium model classroom with computers or taking an online course. Both of these systems feature math lectures online, so you cannot ask questions. This type of delivery system requires a different type of note-taking in order to learn the material.

Warming Up for Math Class

The first step of mental preparation for note-taking involves "warming up" before class begins. Just as an athlete must "warm up" before a game begins, you must "warm up" before taking notes. "Warm up" by reviewing the previous day's notes, reviewing the reading material, reviewing the homework, preparing questions, and working one or two unassigned homework problems. This mental "warm up" before the lecture refreshes your memory, making it easier to learn the new material. Students who "warm up" are "ready to go" as soon as the instructor starts lecturing, and they keep up with what is discussed in class.

How to Become an Active Listener and Learner

In many ways, math class is like a race. Every session follows the same pattern:

1. *On your mark*—sit in your desk.
2. *Get set*—get focused and be ready to take notes when the instructor begins.
3. *Go*—keep a steady pace with note-taking.
4. *Last lap*—stay focused, even if you are tired.

5. *Finish line*—class is over. You have good notes. Time to practice for your next race.

In order to complete this process, you must become an active listener. Active listening means integrating the thinking and learning process with listening and recording information.

First, it involves learning the instructor's lecture style, how he or she indicates when information is important. What information do they tend to leave out? Is it best to keep the book open during lecture? How parallel is the lecture with the book? The pace of the instructor's lecture is important. If it is too slow, you must think of ways to remain focused like finding sample problems in the book. If it is too fast, you must find a way to make short cuts in recording information while still getting the details.

Second, you must be a participant in the class. Ask questions. Listen to other students' questions. Be willing to ask questions for clarification when confused because if you are confused, so is someone else. Questions are another way to force an instructor to repeat information or provide more details. Sometimes it is easy to fall into the frame of mind where you just go to class, get the notes down, planning to learn it later. That is passive listening. No more sitting in the back, slouched in the desk, waiting to write whatever the instructor puts on the board.

What do you think about while in math class? First, listen for the instructor to discuss what you learned when you previewed the chapter. Sometimes it helps to have the book open to the chapter the instructor is talking about. You can compare the information from the lecture with the book.

Second, if the instructor is speeding on, ask a question to help you understand what is being discussed. This also gives a mental break for everyone. Just don't ask too many questions that are irrelevant. Classmates do not appreciate those questions.

Third, as the instructor is explaining how to do something, listen for the explanation of why particular steps are completed. If you do not hear the "why," be brave and ask the instructor. Usually the answer will include the rules, properties or laws that explain why certain steps are taken. This will help you and many of your classmates understand what is going on. Make sure you record this information in your notes.

There will be times when the instructor has to present so much information at a fast pace, all you can do is make sure you get it down. One trick is to sit next to someone who is trying to listen as intently as you. When one of you spaces out, the other one will still be focused in and getting the notes down. At the end of class you can compare notes and fill in each other's gaps. If you can, sit close enough to each other to be able to look at each other's notes in order to fill in the gaps. This is particularly helpful in a class that meets for a long time. The catch is that in these situations, you must review the notes as soon as possible because you didn't have as much time to think about the information in class.

So how do you go about reviewing your notes as soon as possible? I have asked this question to many students and have never accepted the answer that they did not have enough time. If you try hard enough, you can always find enough time to review your notes. Try to find time between your classes or during lunch. Even better, get to your next class early and use the quiet time before class to go over everything you just wrote. Even three to five minutes of review improves learning. This is time incredibly well spent. Not only are you embedding fresh information into your memory, you are actively studying for future exams.

Remember, class time should be considered a valuable study period where you listen, take notes and learn at the same time. One way to learn more in class is to memorize important facts when the instructor is talking about material you already know. Another technique is to repeat back to yourself important concepts right after the instructor says them in class. Using class time to learn math is an efficient learning system.

Activity 4.1 Assessing Your Listening Strengths

Now that you have become an effective listener, it is time to assess what strategies you have been using and which strategies you can use to improve these skills. In the section below, answer the questions and then compare your answers to a classmate's.

1. What 'warm up strategies' have you been using in your math class?

2. With this new information, what new strategies can you use to "warm up" for your math class?

3. What active learning strategies have you been using in your math class?

4. What new active learning strategies can you use in your math class?

5. The progression of a math class is very similar to a race. After each race stage below, put a personal statement starting with "I will" on how you will complete that stage.

   a. On your mark–

   b. Get set–

   c. Go–

   d. Last lap–

   e. Finish line–

How to Become an Effective Classroom Note-Taker

Becoming a good note-taker requires two basic strategies. One strategy is to be specific in detail. In other words, copy the problems down, step by step. The second strategy is to listen for the general principles, general concepts and general ideas and record them in your notes.

Copying from the Board

While taking math notes, you need to copy each and every step of each problem on the board, even though you may already know them. While in the classroom, you might understand each step, but a week later you might not remember unless all the steps were written down. In addition, as you write down each step, you are memorizing it. The major reason for recording every step of a problem is to understand how to do the problems while the instructor is explaining them instead of trying to remember unwritten steps when you are studying on your own. It may seem time consuming; however, it pays off during homework and test time.

There will be times when you will get lost while listening to the lecture. Nevertheless, you should keep taking notes even though you do not understand the problem. This will provide you with a reference point for further study.

Put a question mark (?) by those steps you do not understand. As you take notes on confusing problem steps, skip lines; then go back and fill in information that clarifies your misunderstanding of the steps in question at some later point. Ask your tutor or instructor for help with the uncompleted problem steps and write down the reasons for each step in the space provided.

Also, ask your professor for permission to use your smartphone to take a picture of the board. Because many phones allow panoramic photographs, it is possible to create a single image of the entire board. This method does not, however, take the place of taking proper notes. Only view the images you create as a backup or use them to ensure you copied everything down correctly.

Remember, the goal is to get all the details without writing an essay. A procedure to save time while taking notes from the board is to stop writing complete sentences. Write your main thoughts in phrases. Phrases are easier to jot down and easier to memorize. Another strategy to streamline taking notes off the board is to develop an abbreviation system.

An abbreviation system is your way to reduce long words to shorter versions, which you still can understand. By writing less, you can listen more and have a better understanding of the material. For good examples of proper abbreviations , see the chart printed below.

The Goals of Note-Taking

The goal of note-taking is to take the least amount of notes and get the greatest amount of information on your paper. This could be the opposite of what most instructors have told you. Some instructors tell you to take down everything. This is not necessarily a good note-taking system, since it is very difficult to take precise, specific notes while at the same time understanding the instructor. On the other hand, some instructors ask you to stop taking notes during a lecture and just listen. These instructors know that when explaining a major point you need to listen and not write. If you are doing both, you may not understand the important concept. Let these instructors explain the major point and then record your notes. If you didn't have enough time to take the notes, see the instructor during his/her office hours or get the notes from a friend. Getting notes from another student instead of doing your own notes is not a good idea. Notes from other students are thoughts that pertain to their math knowledge to remind them how to do problems that they don't know how to solve. You might not have the same knowledge as they do, so the notes may not help you.

Abbreviations

E.G.	(for example)
CF.	(compare, remember in context)
N.B.	(note well, this is important)
∴	(therefore)
∵	(because)
⊃	(implies, it follows from this)
>	(greater than)
<	(less than)
=	(equals, is the same)
≠	(does not equal, is not the same)
( )	(parentheses in the margin, around a sentence or group of sentences indicates an important idea)
?	(used to indicate you do not understand the material)
O	(a circle around a word may indicate that you are not familiar with it; look it up)
TQ	(marks important materials likely to be used in an exam)
1, 2, 3, 4	(to indicate a series of facts)
D	(shows disagreement with statement or passage)
REF	(reference)
et al	(and others)
bk	(book)
p	(page)
etc.	(and so forth)
V	(see)
VS	(see above)
SC	(namely)
SQ	(the following)
Comm.	(Commutative)
Dis.	(Distributive)
A.P.A.	(Associative Property of Addition)
A.I.	(Additive Inverse)
I.P.M.	(Identity Property of Multiplication)

It is best to take your own notes and then compare them to a classmate's notes to gain additional information.

What you need to develop is a note-taking system in which you write the least amount possible and get the most information down while still understanding what the instructor is saying. The first step to this system is to know when to take notes.

When to Take Classroom Notes

To become a better note-taker you must know when to take notes. The instructor will give cues that indicate what material is important. Some cues include:

- Presenting usual facts or ideas
- Writing on the board
- Summarizing
- Pausing
- Repeating statements
- Enumerating, such as, "1, 2, 3" or "A, B, C"
- Working several examples of the same type of problem on the blackboard
- Saying, "This is a tricky problem. Most students will miss it."
- Saying, "This is the most difficult step in the problem."
- Saying, "These types of problems will be on the test"— such as different types of word problems.
- Explaining bold-print words
- Saying, "This will be on the test."
- The instructor makes a special point of a particular PowerPoint slide, or the instructor brings up one of many Web-based resources

You must learn the cues your instructor gives indicating important material. If you are in doubt about the importance of the class material, do not hesitate to ask the instructor about its importance. If you are not sure if something is important in the lecture, write it down just in case.

While taking notes you may become confused about math material. At that point, take as many notes as possible, and do not give up on note-taking.

When important cues indicate that this information may be on the test, make sure to put TQ in the margin of your notes. These TQs could be the instructor indicating that this information will be on the test. If an instructor does the same type of problem several times and tells you that you must know how to work these types of problems, it is TQ. These TQs need to be reviewed and learned before each test. List these TQs on a separate page in the back of your notebook to review for the test and final exam.

When to Take Online and Emporium Class Notes

If you are taking an Emporium or online math course, taking notes may become confusing because there may not be formal lectures. The note-taking rules have changed for these courses; however, you still need to take notes. Remember the purpose of note-taking is to provide a resource for working homework problems and preparing for tests. I have worked with hundreds of students in these types of course who did not take notes and failed. Many had successfully completed their homework but did not have any notes or very few notes to review for the major tests. Some cues to take online/Emporium notes include:

1. If the instructor conducts a mini-lecture
2. If the instructor conducts a "pull out" section of students
3. When you ask for help from the class tutor or instructor
4. When you email the instructor or tutor for help on a problem
5. When it takes you three tries to get the homework problem correct
6. When you view the video lecture and don't understand the concept

Activity 4.2 Assessing Your Note-Taking Skills

Now that you have a better understanding of how to become a good note-taker, it is time to do a self-assessment to discover your strengths and those areas in which you need to improve.

1. Show your math notes to a classmate and see if he or she can read and understand the notes. Write down what they said about your notes.

2. List the type of math abbreviations you use in your notes:

3. Make up some math abbreviations you can use for your notes, and write them below:

4. What cues do you need to listen for when you take notes?

5. Based on reading this section, list five additional cues you can use to take notes:

The Seven Steps to Taking Notes in Math Class

Since most students are not court reporters, the key to effective note-taking is to record the fewest words while retaining the greatest amount of information. As you know, it is very difficult to record notes and, at the same time, fully understand your instructor. The "Seven Steps to Math Note-taking" system was developed to decrease the amount of words you write down and to maximize your math learning.

The system consists of three major components. Steps One through Three focus on recording your notes. Steps Four through Six focus on checking yourself to see how much information is retained. This is done by recalling key words and concepts and putting a check mark by misunderstood information. Recalling information is one of the best learning techniques. Step Seven, the third component, is a math glossary.

Before we get into the specifics of these steps, however, let's first explore how to set up your notes page. Great notebooks are organized and legible; they follow and stick to a singular design.

One of the best math note-taking methods is demonstrated in the figure on the next page. The "Modified Three Column Note-Taking Method" allows you to record everything you need to know to solve math problems in an easy-to-follow guide.

To set up this system, do the following on regular notebook paper:

1. *Label the top space* between the note-book ring and the red line, "Key Words."
2. *Label the other side* of the red line, "Ex-amples."
3. *Next, label "Explanations/ Rules"* about four inches from the red line.
4. *Draw a vertical line* between the "Ex-amples" and "Explanations/Rules" sections.
5. *Record the same information* on the next 10 pages. After using this system for 10 pages, you may not need to label each page.

Modified Three Column Note-Taking Method

Key Words/Rules	Examples	Explanations
Solve a linear equation	5(x + 4) + 3(x - 4) = 2(x - 2)	Have to get x on one side of the = and numbers on the other side of the =.
Distributive Property	5x + 20 + 3x - 12 = 2x-4	Multiply numbers to the left of the ( ) by each variable and number in the ( ).
Commutative Property	5x + 3x + 20 - 12 = 2x-4	Regroup numbers and variables.
Combine like terms	8x + 8 = 2x - 4	Add x's together and numbers together .
Additive Inverse Property	8x - 2x + 8 = 2x - 2x -4 6x + 8 = -4	Subtract 2x from both sides to get variables all on the left side of the =.
Additive Inverse Property	6x + 8 - 8 = -4 - 8 6x = -12	Subtract 8 from both sides to get numbers all on the right side of the =.
Multiplicative Inverse Property	$\frac{6x}{6} = \frac{-12}{6}$	Divide both sides by 6 to get x by itself on the left side of the =.
Simplify	x = -2	Solution. Now, check your answer.
	Insert new problem	

Once you have set up your page, follow the seven steps listed below.

Step One – Record each problem step in the "Examples" section.

Step Two – Record the reasons for each step in the "Explanation" section by using abbreviations; short phrases, not sentences; keywords, properties, principles or formulas.

Step Three – Record keywords/concepts/rules in the left two-inch margin either during or immediately after your lecture by reworking your notes.

Step Four – Cover up the "Example and Explanation" sections and recite out loud the meaning of the key words or concepts.

Step Five – Place a check mark by the key words/concepts that you did not know.

Step Six – Review the information that you checked until it is understood.

Step Seven – Develop a math glossary for difficult-to-remember key words and concepts.

After practicing this note-taking system, you may want to modify it to meet your personal note-taking needs. Some students wait and convert their notes into a three column system after class. They also put information from the math book into the three columns. Other students use graph paper and turn it landscape in order to make the columns wider.

In some cases, auditory abstract learners do not take extensive notes. I have seen these students take down only a few notes and spend most of their working memory understanding abstract concepts. For the most part, this is fine. If you are this type of student, however, you should at least take a few notes in order to remind yourself of key concepts. You should also still develop a math glossary. It is very important for you to remember the vocabulary that connects to abstract learning. Knowing these words frees up working memory to solve difficult equations on important tests.

Creating a Math Glossary

The third component in the Seven Steps to Math Note-Taking is devoted to developing a math glossary. Since math is sometimes considered a foreign language, understanding math vocabulary becomes a key to learning math. A math glossary for each chapter dramatically improves learning and allows you to better remember key concepts. Your glossary can be a combination of lecture notes, text readings or text notes (the latter two are discussed in the next chapter).

A good glossary is a key to success for students who have good language skills but have difficulty learning math. Even though math is an abstract subject, it is still learned through using language to recall how to work problems. In fact, some students talk their way through solving equations by using their math vocabulary.

For this reason, a math glossary should be created for each chapter to define math vocabulary words and concepts. Label a section in the back of your notebook "Math Vocabulary for Chapter One." Your glossary should include all words printed in bold print in your text, words emphasized by your instructor and any other words you do not understand. If you are not sure whether or not a word is important, add it anyway just in case.

The glossary should be divided into three areas, which include the book definition, your definition in your own words, and, if appropriate, an example. When you are finished, your glossary will look similar to the note-taking system you learned about in the last section. If you cannot explain the math vocabulary in your own words, ask your instructor or tutor for help. If your instructor or tutor cannot help you, then visit an online dictionary to find definitions. Review websites like math.com for terms from middle school and high school courses, and MathWords.com for terms that range from beginning algebra to college calculus. You can also, of course, simply Google a word for a direct definition.

Some students record vocabulary words with a digital recorder or smartphone. If you use a recorder or phone-app, leave a few seconds between a word and its definition. The definition should be in your own words. These vocabulary words are usually the words that you could not remember from your glossary.

To practice your vocabulary words, play the recording when you have a few minutes between classes (using headphones). This process keeps the words active in your mind until they finally pass into your long-term memory.

Once you listen to the vocabulary word, pause the recording and repeat the definition back to yourself. Replay the recording and listen to the definition. If you did not repeat it back correctly, then continue repeating it until you get it right. Keep practicing until you can correctly repeat back all the definitions. This is an excellent way for auditory math learners to memorize and learn vocabulary. Other types of learners can also use this learning system, because the repetition drills these words into your long-term memory.

Students who are visual or kinesthetic (hands on) learners may be able to learn math vocabulary more effectively by developing a virtual web-based math glossary. StudyStack.com is a free website that is designed to help people memorize and learn information.

Using the StudyStack website, you can use your computer to develop and display a stack of "virtual cards" of information that you want to learn. Just like flashcards, you can review the cards at any time, at your own pace, and you can discard the cards that you already know. The site is set up to find information on any subject, to develop your own flashcards, to play games such as hangman and matching, and also to help you study anytime/anywhere, 24/7. You can also print out the cards and export them to your cellphone, tablet or computer.

Vocabulary Card Examples

Front	Back
Multiplicative Identity Additive Identity	Identity property a times 1 = a Additive identity a + 0= a
Order of polynomials	Place the terms in descending order of exponents. Highest on the left to lowest on the right. x cubed, next x squared, x to the first, constant (number).
Decimals — rational numbers and irrational numbers	Rational numbers have terminating or repeating decimal equivalents. e.g. ¾ = .75, 5/7 =.714285714285... Continuing but non-repeating or nonterminating decimals are called irrational numbers (e.g. the square root of 2 and pi).

Visual math learners can also learn vocabulary the old fashion way — by putting a vocabulary word on one side of a flashcard and the definition on the other side. Auditory math learners can also develop flash cards, making sure to say the information out loud while learning the information.

It is also helpful to write vocabulary words next to the homework problems they are associated with. The examples above are 3x5 cards that show only words. These memory devices can be even more effective if you include pictures or other diagrams.

How to Take Notes for Online and Emporium Model Courses

Modern math students have more course options than any generation in the history of higher education. Two of the newest options, online courses and emporium model courses, differ greatly from classes held in traditional math classrooms.

Online distance courses allow students to use course software to complete a class outside of the classroom. Students are required to finish a pre-set curriculum by a certain date, though they are mostly left to complete the work where and when the want.

Emporium Model courses, on the other hand, are taken in specified classrooms on specified dates. These courses, often held in computer labs, do not typically include lectures, nor do they require students to work at a specific pace. These courses often use similar software to that used in online distance courses, though students typically benefit from having a professor or a group of tutors in the room to answer their questions.

Because there are no planned lectures, most emporium students dive immediately into their homework without taking any notes—this despite the fact that most math software programs allow students to easily find specific passages from the required textbook and even provide extremely detailed, video-based walkthroughs. Having observed many emporium classes, however, I've noticed most students prefer to rely on built in support buttons, which answer their questions as they progress through their assignments. This practice—also common among online distance learners—is extremely dangerous. While these support buttons are often helpful, they are intended to provide reminders, not to take the place of proper lectures.

Just because you are allowed to move at your own pace does not mean that you should skip lectures and move directly to assignments. If you want to succeed in your online or emporium course, you must treat it exactly the same way you would a traditional course. This means that you must read the text and take copious, detailed notes.

With this in mind, online and Emporium courses require their own, unique note-taking system. Traditional classroom notes are mainly taken to help you complete your homework assignments. Online and Emporium notes are taken to help you review for online quizzes and tests. This means that it is more important than ever to remember every step to every type of problem.

Note-Taking Memory Cues and Steps

The first thing you should notice about this note-taking system is that it looks similar to the "Seven Steps to Math Note-Taking" system given in the previous unit. Any similarities between the two systems are all cosmetic. The memory cues and problems steps are much different in the Emporium system.

Using this system enhances your math learning as you complete new units in your software. It also provides you with a set of comprehensive study guides for quizzes and tests. Students can use this note-taking system in two different formats. Some students will use it while doing their homework. Other students will use their regular note-taking system and then transfer their notes into this system after class. I prefer that you use the first way because it will save you time in the long run; however, if you do not have access to the Internet at home, you may have to use the second system and rework your notes when you get back from class. If this is the case for you, you may want to print out or write down the problem steps you find through the "Help" and "Example" buttons in your software. This way you won't be at a disadvantage come test day.

When taking notes in an online or emporium model course, you should focus on any problem that you had to attempt two or three times. Also, focus on any problems on which you required help from your professor, lab assistant or online tutoring program.

Setting up Your Notebook Page

Before you take any notes, you first need to label the top of your notebook page with the homework section and problem numbers the notes refer to. Also, include the date on which you are working on these problems. Once you've done this, complete the following steps:

1. *Label the top* space between the notebook ring and the red line, "Key Words."
2. *Label the other* side of the red line, "Examples."
3. *In the spot* to the right of "Examples," label a third section "Explanations/ Rules."
4. *Draw vertical lines* on each side of the "Examples" section. By doing so, you are setting up three different columns.
5. *Record the same* information on the next 10 pages of your notebook. After using this system for 10 pages, you may not need to label each page.

Seven Steps to Taking Online or Emporium Notes

Once you have set up your page, follow the seven steps listed below:

Step One – Record a problem and each of its steps in the "Examples" section.

Step Two – Record the reasons for each step in the "Explanation" section by using abbreviations; short phrases, not sentences; keywords, properties, principles or formulas. If you do not know the explanations for a particular step, click on your "Help Me Solve This" or "Example" button. Condense this information in sentence(s) above and/or below the problem. If you do not understand the phrases you find in the help sections, click on the "Textbook Help" button and locate the definitions of any words you don't know. If you still have difficulty, click on the "Video" button or ask your instructor or tutor for help.

Step Three – Record key words and rules in the left two-inch margin either during or immediately after an online lecture by reworking your notes. If you are in the middle of an assignment, find these words at the right of each step in the "Help Me Solve" and "Examples" sections.

Step Four – Cover up the "Example and Explanation" sections and recite out loud the meaning of the key words or concepts.

Step Five – Place a check mark by the key words/concepts that you did not know.

Step Six – Review the information that you checked until it is understood.

Step Seven – Develop a math glossary for difficult-to-remember key words and concepts.

When you finish, your notes should look like the notecard printed on the next page. After practicing this system, you may want to modify it to write down only the explanations that you do not understand or feel you might forget. As mentioned earlier, some students wait to convert their notes into a three column system until after class. This is perfectly fine. The key is to organize the details in your notes into a system that connects problem steps with the reasons the steps are necessary.

The Importance of Emailing Your Teachers

Students taking online courses especially need to develop and use this note-taking system. Online students do not have the luxury of taking notes while asking a tutor or instructor to help them solve the problems. Online students must depend more on "Help" and "Example" buttons. If these buttons don't help you understand how to solve a problem, you need to take notes from the support videos and textbook.

Taking all these notes may be time consuming; however, you need to put the information in your own words in order to better understand the concept.

The next logical question is, "What do I do if after taking the notes from various resources, I still don't understand how to solve the problem?"

The answer is very obvious but according to many online instructors most students don't do it. *Email your instructor and ask about how to solve difficult problems.* That is not to say that the instructors don't get emails from their students. The emails they do get are usually technical questions about accessing or operating online homework. Many math instructors wish the questions were about solving homework problems.

Emailing math instructors to help you solve homework problems is similar to using the listening skills we discussed earlier in this chapter. Don't email the instructor the problem just to indicate that you cannot do the problem; send the problem along with your attempted problem steps, and indicate where you got stuck. After the instructor helps you solve the problems, take notes on the steps.

Next, ask the instructor to guide you to online resources for solving similar problems. Check those resources to see if you did review them or if they are new resources.

If you email your instructor a question about a problem, you will probably make his or her day. Conversely, the Instructional Technology team at your college loves to answer technological problems. Direct to them any and all such inquires.

If you are in the Emporium model, don't be afraid to ask your instructor or tutor to help you with notes. Most students have difficulty filling in "Explanation" sections. Have the tutor or instructor explain the reasons for a problem's steps. Don't, however, write down their explanation. Rephrase the explanation into your own words, and ask them if this is what they meant. Then write down your explanation because you remember your own words better while taking tests. This translation helps transfer math concepts into long-term memory and abstract reasoning.

Emporium/Online Three Column Note-Taking Method		
Key Words and Rules	**Examples**	**Explanations**
Words from the right side of problem	The problem	Sentences that describe the next step or sentences at the end of the problem
Solve a linear equation.	5(x + 4) + 3(x - 4) = 2(x - 2)	Get x on one side of the = sign and a number on the on the other side
Distributive Property	5x + 20 + 3x - 12 = 2x - 4	Use the distributive property: multiply number outside () by the variable and number
Add like terms	8x + 8 = 2x - 4	Add numbers and like vari-ables
Additive Inverse	8x - 2x + 8 = 2x - 2x - 4	Get numbers on one side
Subtract like terms	6x +8 = - 4	Get the variable on one side.
Additive Inverse	6x + 8 - 8 = -4 - 8	Get the x on one side and the number on the other
Simplify	6x = - 12	Simplify the equation.
Multiplicative Inverse	$\frac{6x}{6} = \frac{-12}{6}$	Divide by 6 to get x
Simplify	x = -2	Solution. Does it make sense?

Chapter Four: Checklist for Knowledge and Behaviors

Put a check after understanding the knowledge statements and fill in the behavioral activities.

Chapter Knowledge

1. I understand that effective listening is the first step to excellent note-taking. _____
2. I know that the gold of note-taking is to write the least amount possible to record the most information. ____
3. I know my instructors cues for writing down important information. ____
4. I understand that the Modified Three-Colum Note-Taking Example is an excellent example to use for both taking notes and testing myself on the information. I cover up the left side of the note page and recall the information. I study the information I did not recall. ____
5. I know to write down everything the instructor writes on the board or power point slide. ____
6. I know that asking questions in math class can decrease my homework time. _____
7. I know that if I wait too long to review my notes that I may not understand them making it more difficult to learn. _______
8. I know that online and Emporium students need to use a special note-taking system that differs in many ways form traditional note-taking systems. _____
9. I know that when beginning an online or Emporium course, figure out exactly who to contact should there be any technical problems. _____
10. I know that online and Emporium courses require more independent learning and motivation. _____

Behavioral Activities

1. I show up to class at least a few minutes early to "warm up" to take notes. ____
2. I always bring my note book, textbook, and calculator to class._____
3. I sit in the front or middle of the classroom and practice good listening techniques. ___
4. I ask questions about anything that is not clear. _____
5. I only use my smart phone when the instructor allows me to record information or take pictures of the board. _____
6. I review my note right after class and fill in what I missed. ______
7. I compare my notes with a classmate before the next class. ______
8. I have developed a math glossary to help me understand math vocabulary. ____
9. I visit my instructor to review the classroom notes I don't understand. ______
10. I email my instructor about concepts or problems I don't understand. _____

Name: ______________________________ Date: ______________

Assignment for Chapter 4

Answer the following questions on a separate sheet of paper.

1. Review and use the Modified Three-Column Note-Taking figure as a model for your notes. List and define the key words that were discussed while using this system:

2. What are two reasons for warming up for a math class.

3. List and explain three strategies to help you become an active listener.

4. Describe the listening and questioning skills you need to communicate with your instructor when visiting his or her office.

5. List the reasons you need to copy down each step of the math problem.

6. List and explain three ways a math glossary can improve math learning.

7. How is the note-taking system for online or Emporium Model different than classroom notes.

8. Describe the steps to emailing your instructor about the problem you are having difficulty solving.

9. Describe the overall purpose for taking classroom, Emporium or online class notes as it relates to home work, learning and passing tests.

10. List and explain the Seven Steps to Math Note Taking. If you are taking an online or Emporium model class, list the Seven Steps to Taking Online Math Notes instead.

How to Improve Time Management and Reading Techniques

5

In Chapter 5 You Will Learn These Concepts:

- ✓ How to develop a detailed study schedule
- ✓ How to select study goals and target grades
- ✓ When to read your textbook
- ✓ How to effectively read your math textbook
- ✓ How to effectively read an online textbook or chapter

How to Develop a Study Schedule

Before starting to develop a study schedule, let's look at studying and learning effectiveness based on educational psychology. Educational psychologists have conducted research on studying and memory and found out the best time to study is right after class. Research shows that most students lose up to 50% of the information learned in class by the next day. The closer you can schedule doing homework, reviewing your notes or reading your textbook after class, the more likely you'll retain information.

If possible, when working out your schedule, reserve a one or two hour space right after class for studying. If you have back-to-back classes, then schedule study time for as soon as possible after your last class. Even if your schedule has very few breaks, spend at least a few minutes reviewing your notes the same day. Remember, studying right after class is the best way to learn.

Educational psychologists have also conducted research to see what students remember most clearly the next day after studying several subjects. Research indicates that the subject you studied last is what you remember best the next day. This means that you should review your most challenging subject before going to bed each night. This is especially true the night before a test.

Biological Clock

Research also shows that your biological clock has a significant impact on your ability to develop study habits. Most of us learn best at different times of the day.

Try to schedule your study times to match your daily biological clock. If you are a morning person, then don't schedule your study time for late at night. If you are a late-night person, then don't schedule your study time for early in the morning. It will take about two weeks to develop this new schedule into a habit. Once you have, your body and your mind should sync up, giving you the best chance to retain information.

Schedule Weekly Study Time

Now let's look at how you can develop a study schedule while keeping these important points in mind. There are two basic reasons for developing a study schedule: To schedule your study time and to become more efficient at studying. Before moving on, complete Activity 5.1. This way you will know how you are currently scheduling your time.

In order to become a more efficient studier, you need to set aside a certain amount of study time each week. Rather than setting up a number of daily study hours, you should focus on the number of hours, per week, you plan to devote to studying. How many hours do you study per week? Ten hours, 15 hours, 20 hours, 30 hours? Without knowing the amount of your study hours per week, you will not know if you are studying at a productive rate. Many students believe they are studying enough to make their desired grade, when in truth, they are coming up drastically short.

Example: If your goal is to make a B average, and with studying 15 hours per week you make all B's on your tests, then the goal has been met. However, if you study 15 hours per week and make all D's, then you need to increase your study time and/or change your study methods. By monitoring your grades and the number of hours you study per week, you can adjust your study schedule to get the grades you want.

By monitoring your grades and the number of hours you study per week, you can adjust your study schedule accordingly.

Example: You are at the mall on a Sunday afternoon, shopping for clothes, when you start feeling guilty. You have not started studying for that math test on Monday. If you had created a study schedule, you could have arranged to study for the math test on Saturday and still have been able to enjoy the mall on Sunday.

The second reason for developing a study schedule is to use time more efficiently. Efficient study means knowing when you are supposed to study and when you do not have to study. This approach will help keep you from thinking about other things you should be doing when you sit down to study. The reverse is also true. When doing other, more enjoyable things, you will not feel guilty about not studying.

A study schedule should be set up for two reasons: To determine the amount of study time you need, per week, to get the grades you want and to set up peak efficient study times.

How to Prioritize Your Time

To develop a study schedule, review the Planning Use of Daily Time chart printed on page 98. Use the chart to map out planned study times—feel free to make enlarged copies if you wish.

The best way to begin to develop your study schedule is to fill in all the times you cannot study. Do this by following the steps printed on the next few pages. Some of these steps may not apply to you. If so, replace the step with something unique to your schedule.

1. *Fill in all your classes by putting code C.* For example, if you have an 8:00-9:30 class, draw a line through the center of the 9:00 a.m. box on the study schedule.

2. *Fill in the time you work with code W (W = work).* This may be difficult, since some students' work schedules may change during the week. The best way to predict work time is to base it on the time you worked the previous week, unless you are on a rotating shift. Indicate your approximate work times on the study schedule. As your work hours change, revise the study schedule. Remember: Your study schedule should be structured around the number of hours a week you plan to study. Realize that while your work times might change every week, your total weekly work hours usually remain the same.

3. *Decide the amount of time it takes to eat (E = eat) breakfast, lunch and dinner.* This time slot should include both food preparation and clean up. Keep in mind that the amount of time it takes to eat may fluctuate. Eating time also includes time spent in the student cafeteria. If you have an 11:00-12:00 or 1:00-2:00 lunch break, you

might not eat during the entire time; you could be there both socializing and eating. Still put code E in the study schedule, since the main use of your time is for eating.

4. *Include your grooming time (G = grooming).* Some grooming activities include taking a bath, washing your hair or other activities that you do to get ready for school, dates or work. Grooming varies from minutes to hours per day for college students. Mark your study schedule with code G for the usual amount of time spent on grooming. Remember that more time might be spent on grooming during weekends.

5. *Include your tutor time (T = tutor).* This is not considered study time. Tutor time is strictly that which you spend in a tutoring session. If you have a tutor scheduled or meet weekly with your instructor, mark these times with code T in the study schedule.

6. *Reserve time for family responsibilities on the study schedule (F = family responsibilities).* Some family responsibilities include taking your child on errands, mowing the lawn, grocery shopping and taking out the garbage. Also, if you have arranged to take your children some place every Saturday morning, then put it on the study schedule using code F.

7. *Figure out how much time is spent on cleaning each week (CN = cleaning).* This time can include cleaning your dorm room, house, car, and clothes. Cleaning time usually takes several hours a week. Indicate with code CN that you have cleaning time on the study schedule, and make sure it is adequate for the entire week.

8. *Review your sleep patterns for the week (SL = sleep).* Your sleep time will probably be the same from Monday through Friday. On the weekend, you might sleep later during the day and stay up later at night. Be realistic when scheduling your sleep time. If you have been sleeping on Saturday mornings until 10:00 a.m. for the last two or three years, do not plan time at 8:00 a.m. to study.

9. *Figure the amount of weekly social time (SC = social time).* Social time includes being with other people, watching TV or going to church. It can be doing nothing at all or going out and having a good time. You need to have some social time during the week or you will burn out, and you will probably drop out of school. You may last only one semester. If you study and work too hard without some relaxation, you will not last the entire school year. Some daily social time is needed, but do not overdo it.

10. *Figure the amount of travel time to and from work (TR= travel time).* Travel time could be driving to and from college or riding the subway. If possible, use travel to listen to recordings of your class or to review notes. Travel time may vary during different times of the year.

11. *Recall other time obligations that have not been previously mentioned (O = other).* Other time obligations may be aspects of your life, which you do not want to share with other people. Review the study schedule for any other time obligations and mark them.

Once you have finished all of the steps, count up all the blank spaces. Each blank space represents one hour. You might have several half-blank spaces, which each represent one-half hour. Add together the number of blank spaces left and write the total in the oval, which is located on the lower right-hand corner of the study schedule.

Next, figure how many hours you have to study during the week. The rule of thumb is to study approximately two hours per week for each class hour. If you have 12 real class hours (not counting physical education) per week, you should be studying 20-24 hours per week to make A's and B's. Write the amount of time you want to study per week in the square, located in the lower left-hand corner of your study schedule. This is a study "contract" you are making with yourself.

If the number of contracted (square) study hours is less than the number in the oval, then fill in the times you want to study (S = study). First, fill in the best times to study. If there are unmarked spaces, use them as backup study time. Now you have a schedule of the best times to study. On the other hand, if you want to study 15 hours a week and have only 10 hours of space, you have to make a decision. Go back over your study-schedule codes and locate where you can change some times.

If you have a problem locating additional study time, make a priority time list. Take the hours away from the items with the least priority. Complete the study schedule by putting in your best study times. Also, feel free to use one of the many computer-based, phone-based and tablet-based software applications that send you notifications about important events or due dates. These programs, which usually come pre-installed, send you text messages or emails to make sure you don't miss important points on your schedule.

How to Choose the Grade You Want

Determine what grade you want to make in the math course and write the grade on the study schedule. The grade should be an A, a B or a C. Do not write an N, W, X or F because these grades mean you will not successfully complete the course. Do not write D because you may not get credit for the course or be allowed to take the next course. In fact, it is not wise to write C, either, because most students who make a C usually fail the next math course. Your selected grade is now your goal.

Currently, you have a study schedule representing the number of hours of study per week. You also have a course grade goal. After you have been in the course for several weeks and get back the results of your first tests, you will know if you are accomplishing your goal. Should you not meet your goal, improve the quality and quantity of your studying or lower your course grade goal.

How to Create a Weekly Study Plan

By using the information from your completed copy of the Planning Use of Daily Time chart, you will know which time slots are available for study. You can use this information to both develop an effective study plan for the next week and establish weekly study goals.

Each Sunday, use a sheet of paper to record the days of the week from Monday to Sunday. Indicate the best days to study each subject on your Weekly Study Plan. Give first priority to the best times to study math. Therefore you should choose study times that are as close to class time as possible. Be sure to indicate where you will study. Are you going to go to the math lab? Are you going to study in a group at someone's home or workplace? Are you going to study alone at the library?

Once you have indicated in the Weekly Study Goal Sheet your math study times and locations, fill in your study goals for other subjects. If your math class is the last class of the day, schedule your math study time in the first study-time slot. Therefore, you would choose to begin studying your daily math at 3:00 on Monday, Tuesday, Wednesday and Thursday. You will mark the box on the math line for Monday, Tuesday, Wednesday and Thursday with: 3 p.m., daily work, home.

Planning Use of Daily Time

	Monday	Tuesday	Wednesday	Thursday	Friday	Saturday	Sunday
6:00							
7:00							
8:00							
9:00							
10:00							
11:00							
12:00							
1:00							
2:00							
3:00							
4:00							
5:00							
6:00							
7:00							
8:00							
9:00							
10:00							
11:00							
12:00							

CODES: **C**=Class, **W**=Work, **E**=Eating, **G**=Grooming, **T**=Tutor, **F**=Family, **CN**=Cleaning, **SL**=Sleep, **SC**=Social Time, **TR**=Travel, **O**=Other, **S**=Study

Grade goal: _____

Permission is granted to copy and enlarge Figure 11.

Activity 5.1 Assessing Your Current Time Management

Before filling out your study schedule, we need to access your current status of time management, your GPA, and your math grade goals. Complete this short self-assessment to obtain your current status. You may have to guess at some of the questions. This is OK.

1. In the space below, list the GPA your want to have this semester, as well as the reasons you want to achieve this GPA.

GPA __________
Reasons:

2. In the space below, select the math grade you want to achieve, as well as the reasons you want this particular grade.

Math grade __________
Reasons:

3. In the space below, write down the grade you made on your last math test. If you have not had a test yet, list your last math course grade.

Math test grade or last math course grade__________

4. In the space below, list how many hours you spend in tutoring either at a math lab, learning resource center, tutoring center, or private tutoring session, then write down whether or not you believe this is enough time to achieve the grade you want.

Hours a week in math lab/LRC __________
Explain:

5. In the space below, list the amount of hours you spend studying math every week, as well as your total study hours (for all of your courses). Is this enough time to achieve the GPA you want to obtain? If not, what will you do differently?
Math study hours _______
Total study hours _______
Explain:

Math Textbooks and Homework

Knowing how to read a math textbook and complete homework is vital to succeeding in mathematics. Understanding and retaining material from your text makes it much more likely that you will follow lectures and succeed on tests. While this is true in all college courses, having the skill set to read and digest information from a text is more important in math than perhaps any other subject.

Unfortunately, many students struggle with math textbooks, which are far more technical and harder to follow than those used in other courses. Worse, because math instructors know how to read and learn from these books, they often assume that students are capable of figuring out how to solve math problems through assigned readings. The truth is, most students are never taught how to read a math text. This disconnect is at the heart of many students' struggles with the subject.

It is not enough to merely dive into a math textbook. You must develop a system. The same is also true of homework. Most math students do not have a homework system. They begin their homework by going directly to the problems and trying to work them. When they get stuck, they quit. This is not a good homework system. A good system improves success in homework. It also helps your overall understanding of math. Two key components to a good study system deal with using the textbook as a resource and completing homework to learn math well enough to perform during a timed test. This chapter helps you design a system that allows you to learn from your homework and use the math textbook as an invaluable resource for information.

It's also worth noting that more and more math classes are either being taught online or through the Emporium model. These courses require students to read online textbooks and complete online homework problems. Reading online texts requires a new set of reading skills that most students don't have. In most cases, students don't read the online text or section of the text that is paired with the homework problems. They see no value in reading this material. Also most students start doing homework without a homework plan. After several tries, they may get a few problems correct, but they soon forget how exactly they got there.

In order to succeed in these courses, you need to learn how to read online textbooks and complete online homework. First, however, you should assess your current reading system by completing Activity 5.2.

Activity 5.2 Textbook Reading Assessment

This short assessment will help you understand your approach to reading. Many students don't read their textbook (hard copy or online) because they think it will not help them. This is far from the truth. Even students who do read the text may not do it effectively. Complete the self assessment to see where you stand.

1. Math instructors and students discuss when the best time is to read the textbook. Indicate which one is best for you and why?

When do you read your textbook?
_______ Before class
_______ After class
_______ Honestly, I don't read the textbook
Why?

2. When you read your textbook, do you use a system?
_______ Yes
_______ Sometimes
_______ No
How do you read the textbook?

3. Does reading your textbook help you improve math learning?
________ Yes
________A little
_______ No
_______ I don't read the textbook

4. If reading your textbook does not help you learn math, explain why you think this is true. Is there any way reading your text might help improve your math learning and grades? If reading the text does help you learn math, explain why in the same space below.

When to Read Your Textbook

Most students don't read math textbooks because in the past many math textbooks were poorly written. These books were difficult to understand, composed mostly of example problems, and written for instructors instead of students. Now, math textbooks are more student friendly and feature more support materials. When we discuss reading a math textbook, we are also focusing on the additional products that come with the textbook. This includes solution manuals, online texts, problem hints, support websites and tutor sites.

Math instructors and students often debate over the best time to read a math textbook. Some instructors prefer that students read the textbook before coming to class. Other instructors prefer students read the textbook after class. If students read their textbook before going to class, they become familiar with key vocabulary words and concepts. You don't need to understand every bit of material when you read your book before class. Yet, understanding even just a third of the material will free up your mind to understand more material during a lecture. For this reason, I recommend reading ahead two or three sections and putting question marks by material you don't understand. Make a list of the vocabulary words that will most likely appear in a lecture. When your instructor starts discussing the material, have your questions ready and take good notes. When they start discussing a topic you marked, pay special attention. If you do not understand your instructor's explanation, you MUST ask questions. If there is not enough time for questions, make an appointment with your instructor to go over the material. Reading the text before a lecture helps you better understand it and know when to ask questions.

Reading the textbook after a lecture also has its benefits. This method reinforces the material learned in class and further explains misunderstood information. Instructors who believe in this format think they can explain mathematical concepts better than the textbook or don't want you to get confused by its explanations. Reading the textbook after a lecture helps students to remember notes taken during class. The textbook also explains mathematical concepts in different, more understandable ways. In addition, the textbook might cover material that was not discussed in the lecture that will help clarify a concept.

Reading a mathematics textbook before or after the lecture may depend on how well the book is written. It also depends on your cognitive learning style. Try both ways and see which one works best.

10 Steps to Better Understand What You Read

Before practicing the Ten Steps to Understanding Reading Materials, you need to get to know your textbook. Each textbook has its own organizational pattern for presenting information. It is important to understand how the book works.

- Are there learning objectives at the beginning of the chapter that can be used as a checklist to make sure you are learning everything important?
- How are the vocabulary words highlighted? Are they just bold print? Are they in highlighted boxes?
- How are the mathematical rules highlighted?
- Does each section of the chapter have learning objectives?
- Are the homework problems at the end of the chapter arranged according to the chapter sections?

If you can't figure out how your textbook is organized, go to your instructor or a learning center to receive assistance in "figuring the book out." Once you understand how to use your book, then take the following steps and adapt them in a way that they work for you.

There are several appropriate steps in reading a math textbook:

Step One – *Skim the assigned reading material in order to get a general idea of what the chapter is about.* Skimming is an excellent skill to develop. First, it is an excellent way to get familiar with what will be covered in the next class. Second, as you sit down to study your math after the lecture, it helps you to see how all the "pieces of the puzzle" fit together. Third, after you have learned the material, it is a good strategy to use as a quick ten-minute review so that you do not forget what you learned.

Here are the steps I suggest:

- Read the chapter introduction and/or learning objectives and each section summary and/or learning objectives. As you read, try to see the connections between each section. Remember, when you skim, do not try to learn the material; you simply want to get an overview of the assignment.
- As you skim, think about similar math topics that you already know. What do you already know about the chapter objectives?
- As you skim the chapter, circle (using pencil) new words that you do not understand. You can also skim after learning the material, pretending you are a tutor explaining how each ob-

jective connects with one another. You can even explain in your own words what each vocabulary word means as if you were tutoring. This skimming also helps you remember what you have spent so much time learning.

Step Two — *While reading the textbook, highlight the material that is important to you.* However, do not highlight more than 50 percent of a page because the material is not being narrowed down enough for future study. If you are reading the textbook after the lecture, highlight the material that was discussed in the class. Material discussed both in the textbook and lecture usually appears on the test. The purpose for highlighting is to emphasize the important material for future study.

Step Three — *When you get to the examples, go through each step.* If the example skips any steps, make sure you write down each one of those skipped steps in the textbook for better understanding. Later on, when you go back and review, the steps are already filled in. You will understand how each step was completed. Also, by filling in the extra steps, you are starting to over-learn the material for better recall on future tests.

Step Four — *Revise your marks and highlights.* Maybe you marked them the first time while skimming. If you understand them now, erase the marks. If you do not understand the words or concepts, then reread the page or look them up in the glossary. Try not to read any further until you understand all the words and concepts.

Step Five — *Take notes from your math textbook on principles, properties and rules.* Taking meaningful notes after reading a math textbook helps students remember what they have read. These notes also serve as a review sheet for preparing for a test. This note-taking system is different from taking notes in the classroom. These notes become a quicker resource for information while completing homework and are beneficial in learning math vocabulary. The steps are as follows:

- Before reading the math book, label the top of your notebook page with the chapter and section numbers. For example, at the top of the note page put Chapter 2.3 to 2.6.
- Divide the page into three columns. Label the left one "Terms," the middle one "Examples" and the third one "Definitions and Explanations."
- As you read about principles, properties or rules, write down their names, define them, give one or two examples and explain the process.

Terms	Example	Definitions/Explanation
Multiplication principle	$1/3x = -15$ $3(1/3x) = 3\ (-15)$ $x = -45$	Def. = for real numbers a, b, c, with c not = to 0, if a=b then ca=cb Multiply each side of the equation by 3 (this isolates the x). Multiply the other side by 3.

Step Five — *Learn math vocabulary words.* If you do not clearly understand some words, add these words to the note-taking glossary in the back of your notebook. Your glossary will contain the bold print words that you do not understand. You should have the book definition and the definition in your own words. If you have difficulty understanding the bold-print words, ask the instructor for a better explanation. You should know all the words and concepts in your notebook glossary before taking the test.

Step Seven — *If you do not understand the material, follow these eight points, one after the other, until you are comfortable with the information.*

1. Go back to the previous page and reread the information to maintain a train of thought. Make sure you have learned the previous information correctly.
2. Read ahead to the next page to discover

if any additional information better explains the misunderstood material.

3. Locate and review any diagrams, examples or rules that explain the misunderstood material.
4. Read the misunderstood paragraph(s) several times aloud to better understand its meaning.
5. Refer to your math notes for a better explanation of the misunderstood material.
6. Refer to another math textbook, computer software program or DVD that expands the explanation of the misunderstood material.
7. Define exactly what you do not understand and call your study buddy for help.
8. Contact your math tutor or math instructor for help in understanding the material.

Step Eight—*Reflect on what you have read.* Combine what you already know with the new information that you just read. Think about how this new information enhances your math knowledge. Prepare questions for your instructor on the confusing information. Ask those questions at the next class meeting.

Step Nine—*Review your math textbook notes and math glossary several times a week.* Anytime you have a spare five to ten minutes, review your notes or vocabulary words. These short periods of time allow you to study in little chunks and become confident in your understanding of the material. Try reviewing before the math lecture starts, between classes, or at lunch. Over a period of several weeks you will be amazed at how much you can learn. This process can ensure that you know the information before taking a test.

Step Ten—*Write anticipated test questions.* Research has noted that students have about 80 percent accuracy in predicting test questions. Think about what is the most important concept you just read and what problems the instructor could give you that would test the knowledge of that concept. Make up four or five problems and add them to your problem log (this was part of your note-taking system). Indicate that these questions are from reading the textbook. Review the questions and answers before taking the next test. By using this reading technique, you have:

- Narrowed down the important material to be learned,
- Skimmed the textbook to get an overview of the assignment
- Carefully read the material and highlighted the important parts,
- Recorded important information into a three column note-taking system, and,
- Added to your note-taking glossary unknown words or concepts.

Now that you understand the benefits of reading your textbook, you need to decide when to read the text and how not to procrastinate. As mentioned earlier, it is best to read the text before doing your homework. Look at your study schedule and see when you have scheduled your homework time. Do you have time to read the textbook before that time, or do you want to include reading your textbook as part of your homework time?

When I ask students for the reasons they do not read the text, many tell me that they simply had other things to do. In almost every case, these other things were not that important. When we discuss this issue, students usually admit they do not like reading the text, and they use distractions as an excuse to get out of it. In most cases, these students claim that they have read their math texts in the past, and it did not help them learn how to complete problems. What these students don't realize, is that the only reason they experience ineffective study sessions is that they've never been taught how to correctly read a math text.

If you want to pass your math course, this is an absolutely vital skill. You cannot succeed without fully understanding your text.

How to Read an Online Textbook

Reading an online textbook differs a great deal from reading a traditional textbook. The objectives are the same—you still want to have a better understanding of math and eventually transfer information into your long-term memory—but because online texts come in many different formats, these books require much more preparation before reading.

Before reading an online text, you need to become familiar with how the e-book is formatted. Each online text or e-book has its own organizational pattern for presenting information and resources. It is important to understand how the book works. Before reading, you must ask yourself a series of questions.

- Are there learning objectives at the beginning of the chapter or chapter sections that can be used as a checklist?
- How are the vocabulary words highlighted? Are they bold print?
- Are they in highlighted boxes?
- Can you electronically highlight the text?
- Can you mark a page and come back to it?
- How are the mathematical rules highlighted?
- Does the chapter have a vocabulary check that can become your glossary?
- Does the chapter have highlights or review sections with definitions and concepts?
- Are there additional resources to support chapter readings?

If you have difficulty understanding how to use an online textbook, then go to your instructor, learning center or math lab to receive assistance. Understanding how to use the text improves your ability to learn math.

With that out of the way, it is time to get on to the specific steps you need to follow to read on online text. Note that these steps are similar to those given in the previous section for traditional textbooks. These steps, however, take into account the technological aspects that make reading an e-book unique.

The following 10 steps are designed to help you read and digest information from textbooks presented on computer screens, tablets, or other mobile devices.

Step One—*Click or swipe through the assigned reading material or section(s) pertaining to the homework.*

This is a good idea for many reasons. First, doing so gives you a general idea of the material covered in a given section and thereby gives you a leg up on your homework. Second, this gives you a good idea where to turn when you get stuck on a particular homework problem. Third, it makes it easier to review information before taking quizzes or tests.

To skim effectively, stick to the following pattern:

1. Read the chapter introduction or first page of the section.
2. In the chapter introduction, review the learning objectives. As you read, try to see the connections between each section.
3. As you skim the section(s), highlight new words you do not understand. If you are reading on a tablet, many applications allow you to drag your finger over a particular word or line.

Step Two—*Read with concentration.* Now it is time to highlight specific material within a text—not just vocabulary words, but key concepts and formulas. Remember, the purpose for highlighting is to emphasize the important material for future study and review. Click through the pages and use the electronic highlighting method to mark important material.

Step Three—*Take notes from your math chapter section.* For every example, review each step even if you know the steps. If the example skips any steps, mentally tell yourself the step. If you don't know the step ask for help. You will now understand how each step was completed. By mentally going through these steps you are preparing for your homework and are starting to over learn the material for better recall on tests.

Step Four—*Revise your marks and highlights.* Maybe you already marked them while skimming. If you understand them now, remove the highlight. If you do not understand the words or concepts, then reread the page or see if they are in the vocabulary check or chapter highlight.

Step Five—*Take notes from the text.* The notes should only be on the principles, properties, rules and concepts. Taking meaningful notes, either during or after reading the math textbook, helps students remember what they have read. These notes also serve as a review sheet while preparing for a test. This note-taking system is different from taking notes in the classroom.

The steps are as follows:

- Before reading your online math book, label the top of your notebook page with the chapter and section numbers. For example, at the top of the page write "Chapter 1.1—1.3."
- Divide the page into three columns. Label the left one "Terms," the middle one "Examples" and the third one "Definitions/Explanations."
- As you read, or after you read, write down the principles, properties rules or concepts that you do not know, writing the term, example and definition/explanation.

Step Six—*Develop a Math Glossary.* Review the math vocabulary from your notes, and if you do not clearly understand some words, add these words to the note-taking glossary in the back of your notebook. The back of your note-book should be labeled by chapter and section with a list of these vocabulary words.

Step Seven—*Develop a strategy for understanding difficult material.* If you did not understand the material, follow these suggestions to improve learning:

1. Click or swipe back to the previous page and reread the information to maintain a train of thought.

2. Click or swipe ahead to the next page to discover if additional information explains misunderstood material.
3. Locate and review any diagrams, examples or rules that explain the misunderstood material.
4. Read the misunderstood paragraph(s) several times aloud to better understand its meaning.
5. Replay any math videos pertaining to a confusing section.
6. Refer to another math textbook, YouTube videos, or other online resources.
7. Define exactly what you do not understand and contact your study buddy for help.
8. Contact your math tutor or math instructor for help in understanding the material.

Step Eight—*Reflect on what you have just read.* Combine what you already know with new information that you just read. Think about how this information enhances your math knowledge.

Step Nine—*Review your math notes and math glossary.* This review should occur every week. When you have a spare 10 minutes, review your notes or vocabulary.

Step Ten—*Write anticipated test questions.* Research has noted that students have about 80 percent accuracy in predicting test questions. Think about the most important concepts you just read and how those concepts could be turned into test questions.

How to Read a Single Online Chapter Section

Reading a section of an online text involves a condensed version of the steps listed above. In most cases you are reading this section of the chapter because you are having difficulty answering specific homework problems. With this in mind, the purpose of reading this section is mainly to answer the homework problems and prepare for quizzes or tests.

Step One—*Concentrate on important information as you read the section.* Click or swipe through the pages and use the electronic highlighter to mark the information relating to the probllem. If you understand very little information, you should review the video and then reread the section.

Step Two—*Go through examples similar to your problem.* For those examples, review each step and compare those steps that are related to your problem. Write the problem steps next to your problem steps for comparison. Going through example problem steps helps you solve your problem.

Step Three—*Take notes from your math chapter section.* The notes should be related to the principles, properties, rules and concepts of the problem you don't understand. Take the notes the same way as mentioned in the previous section and expand the "Terms" section to include vocabualry words. These notes can be used to help you solve this problem or similar problems or review for quizzes or tests.

Step Four—*Find additional resources.* If you did not understand material, seek out YouTube videos or math-based websites that contain information specific to your problems.

Step Five—*Review your math notes.* This review should occur every week. Review your math notes before reading the next section or homework. This process will help you to become more successful in reading the sections and problem solving.

By using this technique you can:

- Better understand math book sections,
- Answer homework problems,
- Prepare for quizzes and tests,
- and recall information on your quiz or test day.

Chapter Five: Checklist for Knowledge and Behaviors

Put a check after understanding the knowledge statements and fill in the behavioral activities.

Chapter Knowledge

1. I understand that the major reason students give for making poor grades is lack of study time. ____
2. I know that research says that most students lose 50% of the information learned in class by the next day. ____
3. I know that research says that the subject studied last at night is best remember the next day.____
4. I know that my biological clock will tell me the best time to study either morning, afternoon or night. _____
5. I know that I need to focus on the number of hours a week to study math instead of daily study hours. _____
6. I know that I need some social time during the week however to much social time can lead to poor grades. ______
7. I know that reading the math text is more difficult than reading other texts. ______
8. I understand that learning how to correctly read the math text book will improve my learning and understanding the lecture material. ______
9. I use "Ten Steps to Understanding Reading Material" as an excellent way to comprehend the math text. _______
10. I understand that I need to read the single online chapter section before doing the problems.

Behavioral Activities

1. I study math as soon as possible right after class. _____
2. I have competed my Planning Use of Daily Time schedule. ______
3. I am going to study math and do homework ______ hours a week.
4. After taking the next math test I will review my study hours to see if they need to be increased.____
5. I have completed my Weekly Study Plan. _______
6. I have chosen to make a(n) ______ in my math class.
7. I reviewed my textbook to see how it is organized and understand the pattern. _____
8. I take notes when reading my math text book. _____
9. I developed strategies to help understand reading material when I get stuck. ___
10. I developed a math glossary. _____

Name: ______________________________ Date: ______________

Assignment for Chapter 5

Answer the following questions on a separate piece of paper.

1. How should your biological clock affect the times you schedule study time?

2. What are the reasons to monitor the number of hours a week you study math?

3. List two reasons for developing a study schedule.

4. Complete the Planning Use of Daily Time, select your math course grade and select your overall GPA.

5. What should you do if you want to study 15 hours a week but only have 10 hours open on your schedule?

6. List three connections between math textbooks and homework.

7. When is the best time for you to read the math textbook?

8. Explain how taking notes while reading your textbook and developing a math glossary can improve math learning.

9. What are the reasons many students give for not reading the math text book and how can this lower your grade?

10. List and describe the ten steps to reading a math textbook or online math textbook.

How to Improve Your Homework Techniques

6

In Chapter 6 You will learn these concepts:

- ✓ How to effectively complete math homework
- ✓ How to use metacognitive techniques to solve problems
- ✓ How to effectively complete Web-based and online homework
- ✓ How to best use math labs and Learning Resource Centers
- ✓ How to work with a tutor

Math Homework Techniques

Why do math instructors assign homework? That is a good question. I have asked this question to students, math instructors and math department chairs. Some of them knew the answer, but I was surprised when some of them, including the instructors, did not know the answer. Again, let me ask you the question, "Why do instructors assign math homework?"

Yes, you are correct. Homework is given not just to waste time but to have you practice the math problems often enough to understand the mathematical concept that can be put into your abstract reasoning or long-term memory. In other words, you do math homework to remember how to do the problems during the test. However, just memorizing how to do the problems will create difficulty recalling how to do them on the test.

Doing your homework is an excellent time to practice doing problems as if on a test and to understand the mathematical concepts. Completing homework needs to be a learning experience.

Now that you know the reasons to do your homework, I have another question. "Has anyone taught you the best way to do your homework?" Probably not. The next section will answer that question.

Doing your homework can be frustrating or rewarding. Most students jump right into their homework, become frustrated and stop studying. These students usually go directly to the math problems and start working them without any preparation. When they get stuck on one problem, they flip to the back of the textbook for the answer. Then, they either try to work the problem backwards to understand the problem steps, or they just copy down the answer. Other students use Google or a phone application to automatically generate an answer. After getting stuck several times, these students will inevitably quit doing their homework assignment. Their homework becomes a frustrating experience, and they may even quit doing their math homework altogether.

Now that you know how some students do their homework, the next step is reviewing how you do your own homework.

To start the evaluation, complete Activity 6.1 (located on the facing page).

Activity 6.1 Math Homework Assessment

This short self-assessment will help you understand how you approach your homework. Many students don't do all their homework because they believe doing every problem doesn't help them. Even students who do all the homework sometimes don't do it effectively.

1. Check the appropriate area that best describes your usual homework tactics.

   When doing homework:

   ____ I complete the entire assignment.

   ____ I finish some of the assignment.

   ____ Honestly, I don't really do my math homework.

   Does reading the math text help improve your math learning?

   ____ Yes

   ____ A little

   ____ Not at all

   ___ This question doesn't apply to me. I don't do my math homework.

2. If doing your math homework does not help you learn math, then explain why in the empty space below.

3. In the space below, explain how you typically approach doing your math homework:

10 Steps to Doing Your Math Homework

The following 10 steps are designed to help you complete traditional homework assignments from printed textbooks. If you are in an online class, go ahead and turn to the next section, "10 Steps to Doing Your Online or Emporium Math Homework."

Step One — *Review the textbook material that relates to the homework.* You can also use your glossary and textbook notes in your notebook. A proper review will increase the chances of successfully completing your homework. If you get stuck on a problem, you will have a better chance of remembering the location of similar problems. If you do not review prior to doing your homework, you could get stuck and not know where to find help in the textbook.

Step Two — *Review your lecture notes that relate to your homework.* If you could not understand the explanation in the textbook on how to complete the homework assignment, then review your class notes.

Step Three — *Do your homework as neatly as possible.* Doing your homework – organized and neatly – has several benefits. When approaching your instructor about problems with your homework, he or she will be able to understand your previous attempts to solve the problem. The instructor will easily locate the mistakes and show you how to correct the steps without having to decipher your handwriting. Another benefit is that, when you review for midterm or final exams, you can quickly relearn the homework material without having to decipher your own writing.

Example: Problem: $2 (a + 5) = 0$. What property allows you to write the equation as $2a + 10 = 0$? Answer: The distributive property.

Step Four — *Write down every step of every problem.* Even if you can do the step in your head, write it down anyway. This will increase the amount of homework time, but you are over-learning how to solve problems, which improves your memory. Doing every step is an easy way to memorize and understand the material. Another advantage is that when you rework the problems you did wrong, it is easy to review each step to find the mistake.

Step Five — *Understand the reasons for each step and check your answers.* Do not get into the bad habit of memorizing how to do problems without knowing the reasons for each step. Many students are smart enough to memorize procedures required to complete a set of homework problems. However, when slightly different problems are presented on a test, they cannot solve the problems. To avoid this dilemma, keep reminding yourself about the rules, laws, or properties used to solve problems. A good idea is to write each name of the rule, law or property by several of the homework problems. Then, the homework becomes a test review.

> Example: Solve this equation:
> $6x + 5 = 4x + 1$.
>
> The answer is $x = -2$.
>
> Now put this back into the equation:
>
> $6(-2) + 5 = 4(-2) + 1$
>
> Reduces to $-12 + 5 = -7$, which is $-7 = -7$.
>
> If one side of the equation equals the other, you have the correct answer. If not rework the problem.

Once you know the correct reason for going from one step to another in solving a math problem, you can answer any problem requiring that property. Checking your homework answers should be a part of your homework process because it improves your learning and helps you prepare for tests. Check the answers of the problems for which you do not have the solutions. This may be the even-numbered or odd-numbered problems or the problems not answered in the solutions manual. You can also check your answers by substituting the answer back into the equation, doing the opposite function required to answer the question, or using one of the applications listed in Appendix C.

Step Six—*If you do not understand how to do a problem, refer to these points:*

1. Review the textbook material that relates to the problem.
2. Review the lecture notes that relate to the problem.
3. Review any similar problems, diagrams, examples or rules that explain the misunderstood material.
4. Refer to another math textbook, solutions guide, website, phone application, or online video to obtain a better understanding of the material.
5. Call your study buddy.
6. Skip the problem and contact your tutor or math instructor as soon as possible for help.

Step Seven—*Always finish your homework by successfully completing problems.* Even if you get stuck, go back and successfully complete previous problems before quitting. You want to end your homework assignment with feelings of success.

Step Eight—*After finishing your homework assignment, recall to yourself or write down the most important learned concepts.* Recalling this information will increase your ability to learn these new concepts. This information can be placed in your textbook notes, glossary, or lecture notes, whichever you choose.

Step Nine—*Make up notecards containing hard-to-remember problems or concepts.* Notecards are an excellent way to review material for a test. More information on the use of notecards as learning tools is presented later in this chapter.

Step Ten—*Getting behind in math homework is academic suicide.* As mentioned in Chapter 1, math is a sequential learning process. If you get behind, it is difficult to catch up because each topic builds on the next. It would be like going to Spanish class without learning the last set of vocabulary words. The teacher would be talking to you using the new vocabulary, but you would not understand what was being said.

Using Metacognitive Techniques

Metacognition was discussed in the "Assessing and Using Your Math Learning Strengths" chapter and can be applied to textbook and online homework. It is what students are thinking about when solving math problems. This is a self-monitoring process used to develop problem solving steps to find the solution to the problem. Memorization of problem steps instead of understanding the rules and principles to solve the problems leads to passive learning and unsuccessful problem solving. You need a model to follow when solving math problems.

The model of plan, monitor and evaluate is the framework for solving math problems. This model involves asking yourself questions when solving problems. Planning consists of understanding what the problem wants, the strategies to solve the problem, potential obstacles, understanding what information is required, doing the calculations and predicting the outcome. Monitoring is putting the steps in order, keeping one's place, identifying and finding errors, understanding when additional information is needed, knowing when to use another strategy and knowing when you have part of the answer. Evaluating includes knowing if the answer seems right (number sense), putting the answer back into the equation or doing the opposite of the function to see if the answer is correct and measuring the efficiency of the plan and monitoring.

Now that we know the theory about metacognition, let's apply it to solving some math problems. Let's look at the process for solving linear equations.

Solve: $-3(x-6)+2=2(4x-1)$

The first step is making a plan by asking yourself questions about solving linear equations. Some example questions to solve this equation are:

1. Question: What is a linear equation?

Answer: It must have at least one variable and some numerals or variables and an = sign but no exponents greater than one.

Example: $-3(x-6)+2=2(4x-1)$

2. Question: Are there any ( ) ?

Answer: Yes. Then multiply the number or variable in front times everything inside the ( ).

Example: $-3(x-6)+2=2(4x-1)$
$-3x+18+2=8x-2$

3. Question: Are there like terms on the same side of the equation?

Answer: Yes. Then combine the like terms on the same side only.

Example: $-3x+18+2=8x-2$
$-3x+20=8x-2$

4. Question: Are there letters and/or numbers on both sides of the = ?

Answer: Yes. Then put the letters on one side and the numbers on the other side.

Example: $-3x+20=8x-2$
$-3x+20+2=8x-2+2$
$-3x+22=8x$
$3x-3x+22=8x+3x$
$22=11x$

5. Question: Is there a number attached to x?

Answer: Yes. Then divide both sides by the number that is attached to x.

Example: $\frac{22}{11}=\frac{11x}{11}$
$2=x$

6. Question: Does this seem to be the right answer?

Answer: Yes. The answer is about right. It is not too large.

7. Question: How can you find out if it is right?

Answer: Substitute the answer back into the equations and see if one side equals the other side.

Example: $-3(x-6)+2=2(4x-1)$
substitute 2 for x
$-3(2-6)+2=2(4(2)-1)$
$12+2=2(7)$
$14=14$

Question 8: Is this the right answer?

Answer: Yes. One side equals the other side.

The metacongnitive process involves planing, monitoring and evaluating. Step one is part of the plan, steps two through five involve a combination of planning and monitoring. Step six is the evaluation. Use this as a model of asking yourself questions while solving linear equations. Ask your instructor about exceptions to these problems. Now let's look at the steps for solving quadratic equations. What are some of the questions you need to ask yourself to solve $x^2+6x=16$. The first step to planning is to ask yourself questions about quadratic equations. Some example questions to solve this equation are:

1. Question: What is a quadratic equation?

Answer: It has a square term and has an = sign.

Example: $x^2+6x=16$

2. Question: Does one side of the equation equal to 0?

Answer: No. Then move the term from the right side to the left side by using the opposite sign.

Example: $x^2+6x=16$
$x^2+6x-16=16-16$
$x^2+6x-16=0$

3. Question: Can you factor the left side of the equation

Answer: Yes.

Example: $(x+8)(x-2)=0$

4. Question: Do we have two factors?

Answer: Yes.

5. Question: What are the factors?

Answer: (x + 8) and (x - 2)

6. Question: Can you solve each equation?

Answer: Yes.

Example:

$x + 8 = 0$	$x - 2 = 0$
$x + 8 - 8 = 0 - 8$	$x - 2 + 2 = 0 + 2$
$x = -8$	$x = 2$

7. Question: You should have two solutions. Do you?

Answer: Yes. They are -8 and 2.

8. Question A: Does this seem to be the right answer?

Answer: Yes. They are two answers and they are not too large.

Question B: How can you find out if it is right?

Answer: Put the answers back.

Example: $x^2 + 6x = 16$; Solution one is $x = -8$

$64 - 48 = 16$

$16 = 16$

or

$x^2 + 6x = 16$; Solution two is $x = 2$

$4 + 12 = 16$

$16 = 16$

Question C: Are these the right answers?

Answer: Yes. One side equals the other side.

Like the previous problem, step one is part of the plan, steps two and three are a combination of planning and monitoring, while step four is monitoring. Steps five and six are a combination of planning and monitoring. Step seven is monitoring again, while step eight is evaluating. This is a model you can use to help ask yourself questions while solving problems.

You can ask your instructor about exceptions to these problems. Using metacognition to solve math problems will increase your homework and test-taking success. You may ask yourself different questions based on the different types of problems.

These questions may come slowly at first, but they will speed up and become almost automatic, just like your basic multiplication tables. Your math instructors have already achieved this goal and can solve linear and quadratic equations almost at the speed of light. Don't be intimidated. It has taken them many years to accomplish these skills. With practice, some day you can be just as effective.

Metacognition in Class

Practice these metacognitive strategies in the classroom or when watching math support videos. When your instructor puts the problem on the board or you see it in a video, follow the model of plan, monitor and evaluate. Ask yourself questions such as:

- "Do I understand the problem?"
- "What is this problem asking me to do?
- "How can I solve it using planning?"

The next step is to figure out the first step to the problem and see if the instructor writes down that first step. After doing this, move to the monitoring stage and start writing down the first step and anticipating the next steps.

As for the evaluation part of the process, pretend to check the instructor's steps to make sure they are done correctly (sometimes instructors do make mistakes). If you get confused with some of the steps, ask your instructor to clarify. By doing this, you are practicing metacognitive strategies in the classroom the same way you might while doing your homework.

Activity 6.2 Metacognitive Homework

Many students do not assess their thinking process for solving homework problems. This self-assessment will help you understand what you are thinking while working problems. This will allow you to use metacogntive strategies to solve similar problems.

1. Take one of your homework problems and write it down in the space below:

2. In the space below, write down what you are thinking as you solve the problem. In other words, describe your general approach to finding the correct answer:

3. Now write down a second similar homework problem and record what you are thinking while using the metacognitive model of plan, monitor and evaluate.

Homework problem:

Plan:

Model:

Evaluate:

Distance Learning, Emporium Model Courses, and Online Homework

Community colleges and universities are shifting more and more courses to the online and Emporium models. Both of these models require students to become self-learners and self-motivators. Emporium model classrooms require students to complete online homework without listening to lectures, and online courses require students to do just about everything on their own.

Many distance learning students have difficulty in communicating with their instructors and classmates. These students have told me that they do not know how or when to communicate with their instructor. Also, these students did not know that they could communicate with other students for support and help. As a result of this lack of communication, some of these students withdrew from their courses or made a lower grade than they'd intended.

Students in math classroom courses and distance learning math courses are having difficulty developing strategies for completing their required online homework assignments and learning the material. Students are used to doing homework with paper and pencil, not on the computer. Many of these students have told me that they completed the online homework and submitted it, only to realize later on that they have no homework to review for tests. Also, they had no homework to compare with other students or to use as a basis for in-class questions. In short, they did not know how to use online homework to improve their grades and learning. This is something that must be overcome.

These students need a system built specifically for doing online homework. This system needs to be just as detailed, perhaps even more so, than systems built for traditional, classroom-based math courses. With this in mind, over the next few sections, you'll learn:

- 10 Steps to Doing Your Homework (classroom and online).
- Where to find additional learning resources online.
- How to Recall what You Have Learned

10 Steps to Doing Your Online Math Homework

In some courses, all or part of your homework may be assigned to be done online. This is usually the case if you are taking a distance-learning course, a hybrid or blended course, an Emporium model course, or a course where the computer is an integral part of the learning process. In addition, some lecture instructors may assign homework to be done online as an alternative to homework done from the textbook. Whatever the case may be, you will need to develop study skills that can help you improve your online homework success and learning.

Doing your homework online can be a rewarding and enriching experience, but it can also be frustrating if you run into technical problems. Before you take a course that requires you to do homework online, you should ask yourself these questions:

1. Do I have easy access to computer equipment and software, including high-speed Internet access, to run the homework program?
2. Do I have the necessary computer skills to use the computer and software?
3. Am I comfortable doing my homework on the computer?

If you answered yes to all the questions, then you are ready to begin doing homework online! The 10 Steps to Doing Your Homework are still valid, but we will adapt the steps to online homework.

Step One—*Review the textbook material that relates to the homework.* This step is still essential when doing online homework. However, the software you are using may have an online textbook that has additional resources to help you learn concepts. Some programs provide you with lecture videos, animations, audio clips, and interactive exercises that you can practice working as you read through the examples in the multimedia textbook. Take advantage of these resources to help you learn the material.

Step Two—*Review your video lecture notes that relate to the homework.* If you are taking a course where your instructor does not lecture, you need to develop your own set of notes. Many software programs provide you with lecture videos. Watch these lectures and create your own notes as the presenter goes through the material. One big advantage of watching a video lecture is the ability to pause at crucial points and watch a segment over again if you need further review. In a live lecture, this is obviously not possible, and some students find it difficult to take good notes when the instructor moves too fast.

Step Three—*Do your homework as neatly as possible.* Although you will be doing your homework online, it remains important to do your work on paper and keep it in a notebook. Then, if you have a question about the homework, or if you disagree with the answer given by the program, you will be able to refer to your notes when discussing the homework with your instructor.

Step Four—*When doing your homework, write down every step of the problem.* First, write down the problem statement. This is especially important when doing online homework since many programs generate different problems for each student.

If you do not write down the problem, it can be difficult for you or your instructor to figure out what the original problem was. Next, solve the problem and show each and every step. Finally, enter the answer into the program. It is essential that you enter the answer in the correct wording and format. For example, if the program asks you to enter the answer as a fraction, and you enter the answer as a decimal, your answer may be marked wrong.

Step Five—*Understand the reasons for each problem step and check your answers.* Software programs check your answers and give you immediate feedback. This is one of the major advantages of doing homework online. If you get an answer incorrect, use the feedback along with your notes to try and figure out where you made the mistake. The program may also give you another opportunity to work the problem, perhaps with different numbers, and you should take advantage of this option. These programs also reinforce you for trying the problems by giving you positive statements when you get them correct. This makes many students feel good. However, you are not given positive statements for reading the text or watching a video. You can reinforce yourself for completing these tasks by learning more math and making better grades.

Step Six—*If you do not understand how to work the problem, use the resources provided by the software to learn how to do the problem correctly.* For example, you can ask the program to guide you through the solution one step at a time. Use this approach to learn how to solve the problem. You may also be able to view an example that is similar to the problem you are trying to solve. You may also want to review the video lectures, multimedia textbook, or other online resources to help you understand the concepts.

Step Seven—*Always finish your homework by successfully completing problems.* With online homework, it's easier to end your homework session with feelings of success, since you can usually redo problems until you get the correct solution. Many students find that online homework not only helps them understand the concepts more readily, but it also gives them a boost in morale because of the immediate and positive feedback.

Step Eight—*After finishing your homework assignment, recall to yourself or write down the most important learned concepts.* This is still an important step when doing online homework, so resist the temptation to skip this step.

Step Nine—*Make up notecards containing hard-to-remember problems or concepts.* Some programs will provide students with ready-made notecards, and you can print these out and use them as starting points for your own notecards.

Step Ten—*Stay on schedule.* In a distance-learning, hybrid, or computer-aided course, there may not be fixed deadlines for each homework assignment, and students may be allowed to work at a flexible pace. However, if the course must be completed by the end of the term, it is better to pace yourself and complete the work in a timely manner rather than rushing to complete the bulk of the work just before the end of the term. Be sure to complete the online homework assignments in order. Math is still a sequential learning process regardless of how you do your homework!

Activity 6.3 Establishing Study Period Goals

Before beginning your homework, it is important to establish goals for the study period. Do not just grab a coke and chips, sit down and turn to the homework problems (whether in a textbook or in a computer program). Ask yourself this question, "What am I going to do tonight to become more successful in math?" By setting up short-term homework goals and reaching them, you will feel more confident about math. This will greatly improve your motivation to succeed.

Study-period goals are set up either on a time-line basis or an item-line basis. Studying on a time-line basis is studying math for a certain amount of time. Studying by item-line basis means you will study your math until you have completed a certain number of homework problems.

No matter what homework system you use, remember this important rule: Always finish a homework session by understanding a concept or doing a homework problem correctly. Do not end a homework session with a problem you cannot complete. You will lose confidence by thinking about the last problem you didn't complete or understand instead of all the ones you did understand. If you did quit on a problem you could not solve, return and rework problems you have done correctly.

Before your next homework assignment, set up your study period and homework goals by using "I" statements. For example, "I will do homework for an hour, then take a break." Or, "I will record vocabulary words in my glossary."

Goal 1:

Goal 2:

Goal 3:

Goal 4:

How to Use Math Labs, Tutoring Centers, Learning Resource Centers, and the Emporium Model

Learning how to use your math lab, tutoring center, or learning resource center (LRC) can improve your grades. This is also true of those students who learn to thrive in Emporium Model learning centers, which require mandatory attendance for online learning and supplemental help.

Many students are unaware of the tutorial and learning resources offered at their institutions. Some find out about these resources after they are failing, which in most cases is too late. You need to find the location of learning resources and how to utilize them as soon as possible after course registration.

Some colleges and universities have math labs, LRCs, Academic Enrichment Centers, computer labs, Student Support Services, Disabled Student Services or other specialized labs to help students. High schools often offer after-school or private tutoring. You should ask your instructor or counselor where to get help in math. Do not forget to ask your fellow students for recommendations to get help in math. In the meantime, here are a few of the resources available at most math labs.

Computer Programs

Locate the computer program that best goes with your text. It may be the computer software offered by the textbook publisher or commercially bought software. Ask if you can copy it and use it at another location or on your own computer.

Review the other available software programs to find the one that fits your needs. The newer computer programs are more user friendly, but some of the older programs are more effective.

Web-Based Programs

Most book companies have Web-supported materials to support student learning. These support areas could be an online tutor service, call-in tutor service, or extra homework problems and their solutions. Ask your instructor or lab supervisor about these support materials. If your text does not have these support materials, go to the Web and find them yourself.

Apps

When you cannot make it to the tutor lab or math lab, consider trying one of countless phone or tablet applications available for Apple and Android devices. Developers have created hundreds of applications for learning and practicing key math concepts. These applications are no replacement for face-to-face interaction, but they are useful nonetheless.

Tutoring

Most students believe that tutoring is the best learning resource. However, research has shown that the sessions are useless if the tutor is untrained. Try to work with a trained tutor who has had your course. Explain to the tutor your learning style and suggest that he/she tutor you based on your learning style.

Example 1: If you are an auditory learner, then have the tutor orally explain to you how to solve the problem. Then repeat back what the tutor said (in your own words). Use a recorder or phone-app to record the tutor's explanation so you can play it while doing your homework. Make sure the tutor does not just work the problem for you without explaining the reasons for each step.

Example 2: If you are a visual learner, have the tutors write down the steps to solve the problem. Also, write down the reasons for each step or reference text pages for the reasons. If the tutor cannot write it down for you, write down the steps yourself and ask the tutor to review the steps.

Try not to schedule your tutoring sessions around lunch time since it is usually the busiest time of the day. Have your questions ready from your previous homework assignments. Focus on the concepts you do not understand, not just on how to work the problem. The more specific you are about your homework problems, the more tutorial help you will receive. Do not expect miracles! If you tell your tutor, "I have a test in twenty minutes and do not understand anything about chapter six!" – about all the tutor can do is offer to pray for you. However, past experiences have shown that those who have previously helped themselves to tutoring are most likely to be rewarded with good grades.

Practice Tests

Use practice tests to find out what you do not know before the real test. Ask if the math lab/ LRC offers practice tests. Take these practice tests at least two days before the real test. This will give you at least one day to find out how to work the missed problems and another to review for the test. The more realistic practice tests you take, the better you will do on the real test. Make sure practice tests are timed and do not use any of your notes or the text.

Assessment Instruments

Assessment instruments can be used to place you into the correct course, locate your math weakness and help you understand your learning strengths and weaknesses. If you are not sure that you have been placed into the correct course, ask to take a placement test. Being placed into the correct course is a must to pass math. Ask if the lab has diagnostic math tests to locate your weaknesses. Ask about other assessment instruments, which can be used to help improve your learning.

Other helpful items include, manipulatives, models, posters, graphing calculators, YouTube videos and Google searches.

Manipulatives

If a picture is worth a thousand words, then a model is worth a million for the kinesthetic/tactile learner. If you are one of these learners, ask for what may be called "manipulatives" or "3-D models." Manipulatives and models are concrete representations of a concept that you can physically touch. A good example of a manipulative is the Hands On Equation. The

Hands on Equation uses a simulated balance beam with top like objects to represent variables and dice to represent numbers. Your math lab can order the Hands On Equations by going to www.Borenson.com. Some students have made their own manipulative by using magnetic plastic numbers and letters that are put on a metal board from child games.

Students also go to sign shops and purchase the numbers and letters. The students then set up an equation by using the manipulative and moving the numbers and letters around to solve it.

Students also use the letters to represent rules such as the distributive property. You can also use these manipulatives to understand more difficult mathematical problems.

The calculator is another excellent learning tool for students beginning in pre-algebra and developmental algebra. Graphing calculators match the dynamic cognitive and the kinesthetic/ tactile (hands on) learning styles.

> Examples: On the board put down a(b + c) =. Then on the other side of the equation take additional letters and place them to represent ab + ac. Now you have a(b + c) = ab + bc. Now do the steps over again until your learn it. Then put numbers in to represent the letters. Such as 2(3 + 4) = (2) (3) + (2) (4). Now do the multiplication on each side of the equation and you will get 14 = 14.

Use the graphing calculator to see what happens when you add numbers to the equation. You can see the graph move and then understand the effects. This is an excellent way for trial-and-error learners (dynamic) to understand the equations. If you are in high-level courses, ask for connections intended for uploading graphing calculator programs from a computer. It is important to remember, however, some colleges/universities would rather you learn developmental math without using a calculator. For this reason, you should always speak with your professors and counselors before relying on any technological help.

In general, learning math is a lot like learning to ride a bicycle. You can watch someone else do it, but you only learn by trying it yourself. You must believe in yourself and keep at it. Even if you start off wobbly, as long as you keep peddling, you are riding. But if you do not believe in yourself enough to keep peddling, you will fall. In time, you will take off the training wheels and wonder why you ever needed them. Even if you get rusty after a long absence, you will never again need training wheels.

Math is also something you learn by trying it yourself. Others can assist you with techniques, but in order to make it stick, you need to learn math on your own. As long as you keep trying, you are learning to think mathematically, and you will be able to do it. In the future, even after a long time away from math, you will remember that you were able to master math before and with a little review you still can.

To find other math manipulatives online, click on the "Student Resources" section on AcademicSuccess.com. The website lists numerous helpful resources, many of which are free.

Quick Section Review

- *Math labs and learning centers* are great resources for students who struggle with math.

- *When seeking a tutor,* it is important to find somebody who shares your learning style. This way, you maximize your learning and avoid wasting time.

- *When possible, kinesthetic learners should seek out manipulatives.* Manipulatives help these students grasp key concepts and ideas.

Activity 6.4 Establishing Math Lab, LRC, Tutoring Center, and Distance Learning Resources

Many students are unfamiliar with the math lab, LRC, and distance learning resources provided by their colleges. Students must locate and learn how to use additional resources to improve grades. Complete the survey and resource lists printed below, then summarize the resources you have used or will use.

1. In the space below, list the location(s) and hours of your college's math lab or LRC (there may be more than one).

2. List the tutor times and names that match your course and schedule.

3. List the times and locations for any ongoing group learning sessions. This includes Supplemental Instruction, Peer Assistance Learning, and group study sessions for your course.

4. List the location and times for any support workshops on subjects such as test anxiety, test-taking, test review and final exam review.

5. List and describe three apps you have used or will use to improve your learning.

6. List and describe two distance learning resources you can use for any course.
Summarize your resources:

Distance Learning Resources

For many students, additional outside resources are a necessity for doing well in a course. These outside resources supplement online instruction and in some cases may become primary instruction. The learning resources be either traditional or non-traditional. Traditional learning resources include textbooks, private tutoring, tutor centers, adult education centers, libraries, study skills training, DVDs, and commercial computer programs. Non-traditional resources include call-in tutor centers, online tutor centers, commercial online tutor centers, websites, applications, or Google. These resources often enhance your learning.

Learning resources can also be categorized based on location. There are campus-based, community-based and personal resources. Campus-based resources are becoming very important because of the unexpected number of students who take both distance-learning courses and regular classroom courses. Students who are distance learners and cannot visit the campus look for local community resources to support their learning. Personal resources, such as tablets, have tremendously expanded because of the need to support distance-learning students. Taking advantage of these resources, especially early in the term, can increase your learning and grades. Try to find out what academic resources are available before enrolling in a distance-learning course in math. It doesn't have to be just you and the course. You can always get additional help. Distance-learning students who have access to their campus resources may want to go to the college or university website. This website should give you the resource information you need to make further contacts. You may want to contact the math lab, learning center, assessment center, library, returning adult center or any other place that may be able to assist in your learning. You want to ask them the following questions:

1. Is tutoring by appointment or is it drop-in?
2. When is the best time to get tutoring?
3. Do you have evening hours for tutoring?
4. Do you have any computer programs to assess my study skills and learning

Tutoring with trained tutors can be a great benefit to your learning. Some of these tutors will come to your home. You can find tutors in the phonebook by calling high school and college math departments or by asking your online instructor.

Make sure that the tutors are trained or they have references that you can call. Tutors can be a great help in improving your learning, but most students wait too long to use them. As soon as you are in trouble, or if you anticipate the need for assistance, find it immediately. Don't wait until you are failing.

Importance of Time Management

Students often enroll in Distance Learning Courses because they believe they are easier. Unfortunately, this is not true. Distance Learning Courses require a greater degree of independence and a better sense of time management. The main reasons students fail online or distance courses are related to organization and procrastination. Learning styles and anxiety also play a role, but the ability to keep to a schedule is by far the most important factor when learning how to thrive in a distance-learning environment. For more on how to incorporate proper learning and study strategies within an online study system, visit Emathready.com.

Internet Resources

Online mini-lessons, algebra models, worksheets, virtual algebra solutions, and sites that solve math problems are on certain websites. Some websites are free. Others require a registration fee. Some of these sites have advertisements that pay for the site. You might want to review the sites under Student Math Practice and Learning Sites at AcademicSuccess.com—Student Resources. The sites are arranged by math course name with supplemental learning in calculator skills, virtual flash cards and video lessons.

Using these online sources can improve your math learning and skills. Distance learning students can also find free sites that will solve their submitted math problems.

Additional learning resources can make the difference in being successful in a distance-learning course. Explore the different resources to see which ones meet your learning style and time constraints. If you are not sure which learning resource meets your needs, then try several of the different learning resources to discover which one best improves your learning.

See Appendix C for additional sources.

Using Google and YouTube to Solve Problems

Another way to use the Internet to solve homework problems is to Google the type of homework problem or to look for that type of problem on YouTube.

To Google the problem, you need to define the type of problem. Do this by turning to the textbook or homework section you are working on and looking for the heading. For example, if the section is titled "Dividing a Polynomial by a Monomial," enter the phrase exactly in the search bar. Doing this usually brings up numerous helpful websites. You may have to click on serveral before you find the best resource for your particular learning style, so do not focus on the first link you find unless it is immediately useful.

YouTube is also a great option. The website features thousands of math videos, which usually show the step-by-step processes required to solve specific types of math problems. For many visual and auditory learners, this is the best way to learn. Watching an expert solve a problem and listening to him or her describe the steps allows these students to grasp concepts much more quickly. If you are one of these students, don't hesitate to facilitate your reading sessions with any of the resources described above.

Once you discover which resources or videos work best for you, make sure to save them to your bookmarks. This will make it extremely easy to return to them in the future. This way, these Internet resources will become a part of your textbook reading process.

How to Work Effectively with a Tutor

Redesigned math courses require more independent learning outside the classroom. This means that students now need to begin tutoring the first week of class to review forgotten or unlearned prerequisite math skills. Different students react to this tutoring in different ways. Some claim tutoring only futher confuses them. Others indicate that they could not pass without it. What explains this disparity? Some students have learned how to become a better tutoree and others have not. Effective tutoring requires preparation before sessions and continues as a team approach, which requires student and tutor effort. Take the following steps, and you should be OK.

1. The best time for tutoring is right after class.
2. Schedule tutoring at least three times a week and put it in your phone's calendar.
3. Ask for permission to take pictures of difficult problems.
4. Ask for permission to record explanations of difficult problems.
5. Take effective notes, leaving blank lines between problem steps you don't understand.
6. During tutoring and after, understand that the person holding the pencil is the one doing the learning.
7. Show tutors your notes so that they can see how the instructor solves problems.
8. Let your tutor know your learning style.
9. Suggest that the tutor teach you in your preferred learning style.
10. Visual learners need color coded numbers and variables to aid memory.
11. Show the tutor pictures of difficulty problems from class.
12. Take pictures of tutored problem steps for review.
13. Auditory learners should play back recorded explanations of difficult problems.
14. Auditory learners should also ask the tutor to explain how to work problems.
15. Use "talk out loud" strategies to tell your tutor exactly what you are thinking.
16. Have your tutor show you how to use math Apps to improve learning.
17. Once you find tutors that match your learning style, stay with them.
18. Finish tutoring sessions with problems you can complete.

Now that you have these strategies, you can become a more effective learner and improve your grades. All learners can use these multimodality learning strategies. Start off by implementing a few strategies and continue to other strategies to discover which ones are best for you.

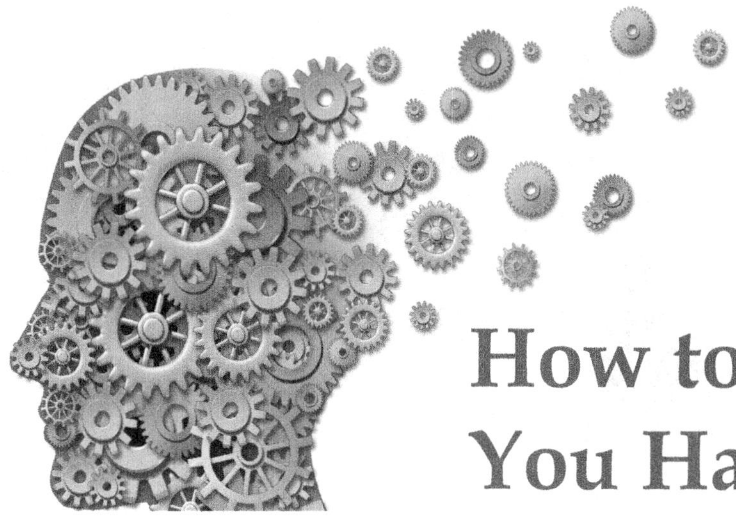

How to Recall What You Have Learned

After completing your homework problems, a good visual learning technique is to make notecards. Notecards are 3x5 index cards on which you place information that is difficult to learn or material you think will be on the test.

On the front of the notecard, write a math problem or information that you need to know. Color code important information in red or blue. On the back of the notecard, write how to work the problem or give an explanation of important information. Make notecards on important information you might forget. Every time you have five spare minutes, pull out your notecards and review them. Glance at the front of the card, repeat to yourself the answer and check yourself with the back of the card. If you are correct and know the information on a card, do not put it back in the deck. Mix up the cards you do not know and pick another card to test yourself. Keep doing this until there are no cards left that you do not know.

If you are an auditory learner, use a digital recorder or phone application just like the notecards. Record important information like you would on the front of a notecard, then leave a blank space on the recording. Record the answer. Play the audio file back. When you hear the silence, pause the recording, then say the answer out loud to yourself. Un-pause the recording and see if you were correct. You can use this technique in the car while driving to college or work.

Review What You Have Learned

After finishing your homework, close the textbook and try to remember what you have learned. Ask yourself these questions, "What major concepts did I learn tonight?" or "What test questions might the instructor ask on this material?"

Recall for about three to four minutes the major points of the assignment, especially the areas you had difficulty understanding. Write down questions for the instructor or tutor. Since most forgetting occurs right after learning the material, this short review will help you retain the new material.

Another great time to review your notes is right after completing an online homework assignment. Make sure you know what section you finish to keep track of your progress. Look at the key words and repeat them to yourself or write them down on flash cards. Review the flash cards as often as you can, and write down any websites that you find helpful, including links to YouTube videos.

When you go home, open up these sites on your computer for a second review, then bookmark them so you can use them in the future. This is an especially good strategy if you do not have a math textbook or your textbook is online. In this scenario, your online bookmarks work the same way a real bookmark would in a tradtional text.

Communicating with Your Instructor and Classmates

Communicating with your instructor and classmates from a distance can be a challenge. The syllabus should have the email address and telephone number of your instructor. The first chance you get, send a "test" email to your instructor to make sure your have the correct address. Some students feel that it is an inconvenience to email their instructors questions. This is not so. The instructor expects your emails and will answer them. However, it is best not to start your email with, "I hate to bother you but…."or "I know this sounds like a stupid question but..." These types of introductions show that you are a novice distance learner and sometimes irritate the instructor. Start your email off with a direct question and you will get a direct answer.

Other students assume that their instructors are on the computer 24 hours a day, seven days a week. This is not true. These students expect the instructors to answer their questions immediately even if it is 1:00 a.m. Usually your instructors have set times of day that they answer email questions. You may want to email your instructor and ask when he or she usually answers the questions and how long it usually takes.

If you have detailed questions, it might be better to have them answered over the phone. Email your instructor and set up a time to call. Make sure you have the questions ready and can record the answers. Your time and the instructor's time is valuable. Even with the best time management, you may not be able to finish the course on time. How do you communicate to your instructor that you may not finish the course on time? DON'T WAIT! Instructors hate to see an email the last week of class saying that you will not finish.

Many distance-learning courses have group chat rooms. These chat rooms are live or the messages can be posted. The instant message capability of many email systems can set the stage for study groups. Ask your instructor if a time is set up for a group chat or study group. Also ask if your instructor will join the chat group and at what times. You may also get the email address of some of the other students so you can keep up with assignments and help each other with difficult problems.

As with any online endeavor, it is important to remember that everything you write, whether on a message board or on Facebook is permanent. If you post something on a social media platform, you must assume your professor will wind up seeing it. This means that you should NEVER post anything negative about your instructor nor complain about a bad test score. I have seen students who tweet something about a classmate or a teacher wind up getting in trouble for what they thought were harmless comments. When writing about school-related matters, always make sure to use appropriate language.

Chapter Six: Checklist for Knowledge and Behaviors

Put a check after understanding the knowledge statements and fill in the behavioral activities.

Chapter Knowledge

1. I understand the reasons instructors assign homework. ____
2. I understand that without a homework system students can become frustrated when doing their home. ____
3. I know that following the 10 Steps to Doing Your Math Homework until becoming comfortable using them as part of my study routine. Then I can adjust the ten steps to make my own efficient system. _____
4. I understand that getting behind in math homework is a major problem and leads to failure. _____
5. I understand that I need to take time after competing my homework to recall what I have learned . This will help me retain the new information in long-term and abstract reasoning. _____
6. I know that the metacognitive model for homework has three steps: plan, monitor and evaluate. _____
7. I know that applying the 10 Steps for Doing Online Homework to distance learning courses or required online homework can increase my math learning. ____
8. I understand that learning how to use math lab/LRC resources that match my learning modality can enhance learning. _____
9. I know to use practice tests at least two day before the test to find out what I don't know before the test instead during the test. ______
10. I understand that distance learning courses are usually more difficult that classroom courses and require more time management. ______

Behavioral Activities

1. I read the related textbook section or reviewed my notes before starting my homework. _______
2. I immediately after class work several homework problems to make sure I understand the skills. ______
3. I work all the homework problems by the due date. _____
4. I use the Metacognitive homework techniques to solve problems. ____
5. On difficult problems I develop a series of questions to ask myself. ____
6. I know review material to help keep it in long-term memory. _____
7. I properly communicate with my instructor to ask to ask homework questions. ____
8. I developed a math glossary. _____

Name: ______________________________ Date: ______________

Assignment for Chapter 6

Answer the following questions on a separate piece of paper.

1. What are the reasons for you to complete your math homework?

2. What should you do before starting your math homework?

3. What are the reasons for writing down every problem step while doing your homework?

4. List and describe the 10 Steps for Doing Your Math Homework or Online Homework.

5. Describe how you can use metacognitive strategies in the classroom or online course?

6. What are your best math lab or Learning Resource Center resources?

7. List and describe how you use three of the apps in Appendix C to improve your tutor sessions and math lab learning.

8. List and describe five resources you can use to improve online learning.

9. List and describe three strategies to help you recall what you have learned.

10. List how you are communicating with your classmates and instructor and some problems you could have with electronic communication.

How to Improve Your Math Test-Taking Skills

7

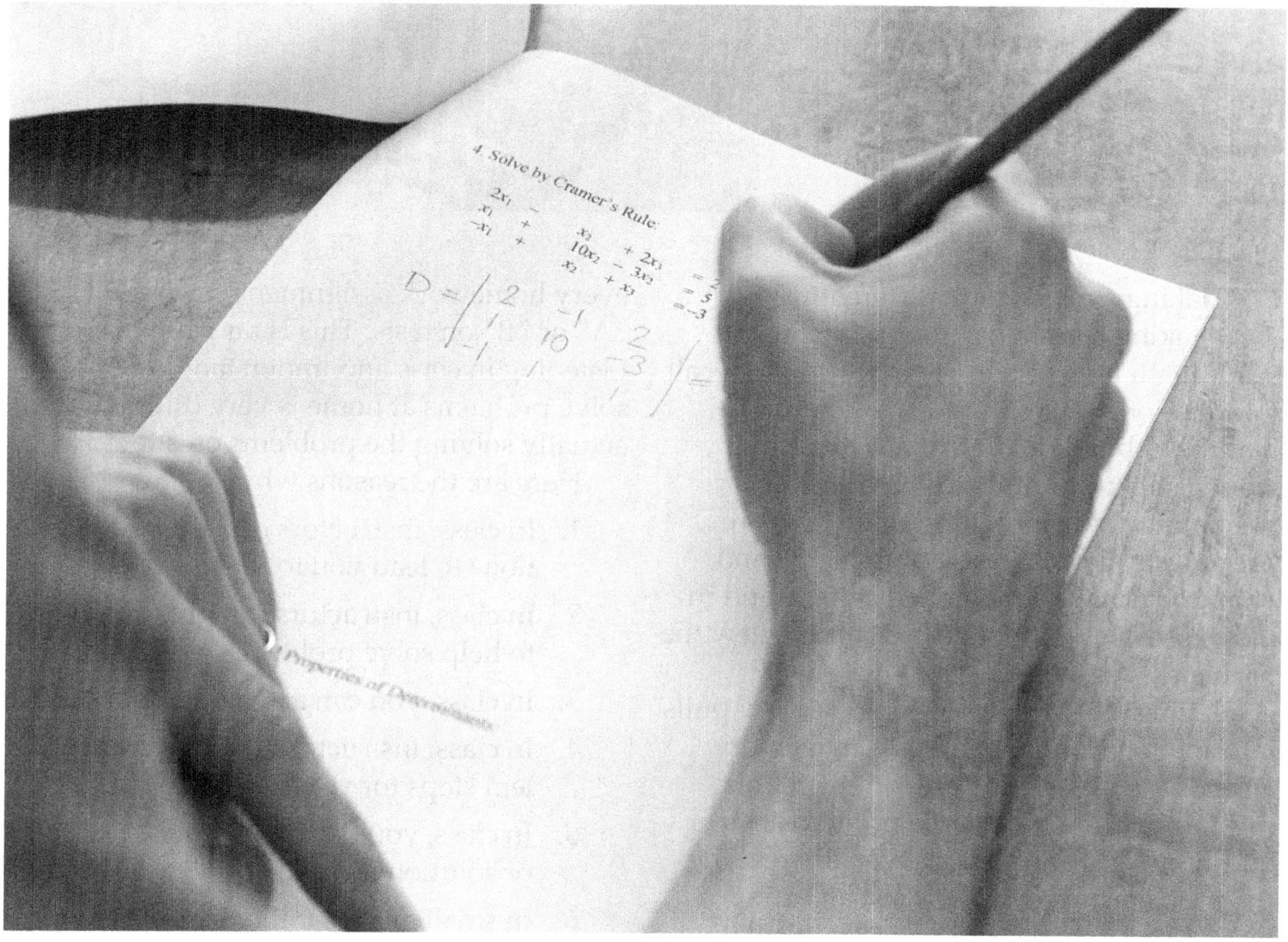

In Chapter 7 You will learn these concepts:

- ✓ Reasons why attending class is not enough to pass tests
- ✓ The general rules to follow before a test
- ✓ Ten steps to better classroom test-taking
- ✓ Ten steps to better test-taking in Emporium and online courses
- ✓ How to analyze old tests for errors and improvement

Why Doing Your Homework and Attending Class May Not Be Enough to Pass

Taking a math test is different from taking tests in other subjects. First, math tests not only require you to recall information, but apply it as well. Second, math tests build on each other, requiring students to know and recall material from previous units. Third, most math tests are speed tests. Fourth, Emporium model and online tests have strict time restraints and in most cases you cannot go back and review the problems.

During a test, your working memory pulls information from your long-term memory and/or fluid reasoning areas. It then puts this information together in order to answer questions. At the same time you are reading the problems, your brain is also deciding how to progress through the test, calculating answers and performing other functions. All this work requires working memory space, so you need to develop test-taking strategies and use accommodations if appropriate to free up enough working memory to correctly solve problems. Remember, math test preparation and test-taking skills are different from those needed for tests in other subjects. You need to have a test-taking plan to demonstrate your total knowledge. Students with these plans make better grades than their peers.

Most students and some instructors believe that attending class and completing every homework assignment ensures an "A" or "B" on tests. This is far from true. Listening in class and understanding how to solve problems at home is very different from actually solving the problems on a test.

Here are the reasons why:

1. In class, instructors can ask you questions to lead you to the next step.
2. In class, instructors can give you hints to help solve problems.
3. In class, you can ask helpful questions.
4. In class, instructors can do all the problem steps for you.
5. In class, you can refer to your textbook or solution manual.
6. In small groups, other students can help you solve problems.
7. There is less anxiety solving problems in class than during a test.
8. If you do not learn how to solve a problem in class, you can go to the instructor or tutor afterwards.
9. While in an Emporium model classroom, homework management systems feature help buttons, which refer you to text sections and videos.
10. While in an online class, you can email the instructor or Google results.

Many students get a false sense of security because they know how to do the math problems while in class or when doing online homework. Then, when the test comes around, they have difficulty solving the same problems. Don't let this happen to you! I have worked with hundreds of students who falsely assume that they understand how to solve math problems because they got all the homework problems correct weeks before taking a test. These students guessed their way into failure, because they did not truly understand important math concepts. They did not take any notes on the problems they missed to review before the test. This reviewing could have moved the information from working memory to long-term memory or fluid reasoning. Correctly answering homework problems is very different from taking tests:

1. While doing homework there is little anxiety.
2. You are not under a time constraint while doing your homework.
3. If you get stuck on a homework problem, you have your textbook and notes.
4. Once you learn how to do several problems in a homework assignment, the rest are similar.
5. In doing homework, you have the answers to at least half the problems in the back of the text and can use apps to solve problems.
6. While doing homework, you have time to figure out how to correctly use your calculator.
7. When doing homework, you can call your study buddy or ask a tutor for help.
8. When doing your homework, you can go to the Web and find online tutoring.
9. While in a Emporium model classroom, the homework management program gives you hints, examples and answers.
10. While in an online class, you can access online live tutor support programs, such as Smart Thinking.

General Pre-Test Rules

1. *Get a good night's sleep before taking a test.* If you cram all night and imagine you will perform well on your test with three to four hours of sleep, you are wrong. It is better to get seven or eight hours sleep and be fresh enough to use your memory to recall information needed to answer the questions.
2. *Start studying for the test at least three days ahead of time.* Make sure you take a practice test to find out what you do not know. Review and work the problems in your problem log. If you did not do a problem log, use the chapter review test to find out what you don't know. An even better way is to work with classmates to make up test questions. Review the concept errors you made on the last test. Meet with your instructor or tutor for help on those questions you cannot solve.
3. *Only review already-learned material the night before a test.* This prevents you from forgetting information. Studying new material sometimes forces learned concepts out of your memory.
4. *Make sure you know all the information on your mental cheat sheet.* Review your notebook to make sure you understand the concepts. Work a few problems and recall the information on your mental cheat sheet right before you go to bed. Go directly to bed; do not watch television, listen to the radio or visit Facebook. While you are asleep, your mind works on and remembers the last thing you did before going to bed.
5. *Get up in the morning at your usual time and review your notes and problem log.* Do not do any new problems. Make sure your calculator is working.

Ten Steps to Better Classroom Math Test-Taking

Most students know how to take a multiple choice test, but have no clue about the best way to take a math test. When taking any test, you are being measured in two areas: the content of the material and how well you take a test. To take a math test you need a plan. This plan will make you more efficient and confident. The plan needs to be developed before you take the test and needs to be followed during the test. I have known hundreds of students who have lost tests points because they did not have a test plan and did not practice that plan before taking the test. Make sure you develop and memorize your mental cheat sheet so you can write down the information on the test (don't use the sheet during the test). Finally, make sure to practice the test-taking steps printed below.

10 Steps to Taking a Math Test (in class)

The following few pages describe a math test-taking method that has help thousands of students thrive in their math courses. Before reading the steps, however, you need to assess your current test taking method. To do this, complete activity 7.1, which is located on page 141. If you are currently taking an online math course, fill out the same form and move on to the "10 Steps to Better Online Test-Taking" section in this chapter.

Step One — *Use a memory data dump.* When you get your test, turn it over and write down the information that you put on your mental cheat sheet. Your mental cheat sheet has now turned into a mental list and writing down this information is not cheating. Do not put your name on the test, do not skim it, just turn it over and write down those facts, figures and formulas from your mental cheat sheet or other information you might not remember during the test. This is called your first memory data dump. The data dump provides memory cues for test questions.

Step Two — *Preview the test.* Put your name on the test and start previewing. Look through the entire test to find different types of problems and their point values.

Put a mark by the questions that you can do without thinking. These are the questions that you will solve first.

Step Three — *Do a second memory data dump.* The second data dump is for writing down material that was jarred from your memory while previewing the test. Write this information on the back of the test.

Step Four — *Develop a test progress schedule.* When you begin setting up a test schedule, determine the point value for each question. You might have some test questions that are worth more points than others. In some tests, word problems are worth five points and other questions might be worth two or three points. Just like in a video game, you must decide the best way to get the most points in the least amount of time. This might mean working the questions worth two to three points first and leaving the more difficult word problems for last. Decide how many problems should be completed halfway through the test. You should have more than half the problems completed by that time.

Step Five — *Answer the easiest problems first.* Solve, in order, the problems you marked while previewing the test. Then, review the answers to see if they make sense. Start working through the test as fast as you can while being accurate. Answers should be reasonable. For example, the answer to a problem of trying to find the area of a rectangle cannot be negative, and the answer to a land-rate-distance problem cannot be 1,000 miles per hour. Clearly write down each step to get partial credit, even if you end up missing the problem. In most math tests, the easier problems are near the beginning of the first page; you need to answer them efficiently and quickly. Completing several easy problems first boosts your confidence, decreases anxiety and guarantees test points.

Step Six — *Complete the difficult problems you know how to do.* These are the problems that have quite a few steps and take some time to complete, but are also problems you can complete without any real trouble. These problems are usually worth more points than the rest of the test, so it is important to finish them before you attempt problems that you don't know how to do. By finishing these problems early on in your test time, you can relax, knowing that you already have quite a few points in the bag. This will boost your confidence as you head into the problems you might not know how to do.

Step Seven — *Tackle the toughest problems.* These are the problems that you might remember a few things about, but you don't know enough to complete them. You might remember the first step, but then draw a blank on what to do next. It is important to begin the problem, even if you know that you cannot complete it. By starting the problem and writing down what you know, you are warming up your brain, which might trigger the "Ah ha!" response. The "Ah ha!" response is when your brain suddenly remembers how to complete a problem. A memory from your homework or a lecture sometimes will pop back into your head right when you need it most. Take some time on these problems, but move on if you are completely stuck.

Step Eight — *Guess at the remaining problems.* Do as much work as you can on each problem, even if it is just writing down the first step. If you cannot write down the first step, rewrite the problem. Also remember that the way you learned how to solve a problem was by writing, not just looking at the problem. Sometimes rewriting the problem jars your memory enough to do the first step or the entire problem. If you leave the problem blank, you will get a zero. You MUST put something down, even if it doesn't make sense. Your instructor may give you points just for the attempt.

Step Nine — *Review the test.* Look for careless errors or other errors you may have made. Students usually lose two to five test points on errors that could have been caught in review. Do not talk yourself out of an answer just because it may not look right. This often happens when an answer does not come out even. It is possible for the answer to be a fraction or a decimal. Remember, answers to math problems do not have "dress codes." Research shows that the odds of changing a right answer to a wrong one are much greater than the odds of changing a wrong answer to a right one.

Step Ten—*Use all of your alloted test time.* Review each problem by substituting the answer back into the equation or doing the opposite function required to answer the question. If you cannot check the problem in these ways, rework the problem on a separate sheet of paper and compare the answers. Do not leave the classroom unless you have reviewed each problem two times or until the bell rings.

Even though we encourage students to work until the end of the test period, most students leave the classroom before the end of the period. These students state that even though they know they should use all their test time, they cannot stay in the room until the end of the test time. After talking to hundreds of these students, I discovered two different themes for leaving the classroom early. First, test anxiety gets so overwhelming that they cannot stay in the room. The relief from the test anxiety (leaving the room) is worth more than getting a better grade. The other reason for leaving the test early is that they do not want to be the last or one of the last few students to turn in their tests. They still believe that students who turn their tests in last are "dumb and stupid." These students also believe that students who turn their tests in first make "A's" and "B's" and those students who turn their tests in last make "D's" and "F's."

These students need a new strategy for the middle or last part of a test. This involves performing a second relaxation technique either in the middle of the test or before reviewing the problems. Students who do a short-term relaxation technique, such as visual imagery, prevent their anxiety from getting so high that they have to leave the room. Other students may do a cognitive technique, such as telling themselves to calm down, stay in the room and finish checking their answers. Students who have extremely high test anxiety can do these techniques during the middle of the test and at the end of the test to take control of their anxiety.

If you are one of these students, you don't need to care about what other students think about you (it's usually wrong anyway). YOU need to fight the urge to leave early and use all the test time.

Remember, passing mathematics is the best way to get a high paying job and support yourself or your family. Do not worry about what others think. DO IT NOW!

Benefits to Turning in Your Scratch Paper

Stapling your scratch paper to a math test when handing it in has several advantages. First, if you incorrectly copy the answer from the scratch paper, you might still get partial credit for your answers.

Second, if you get the answer incorrect due to a careless error, your work on the scratch paper may salvage you a few points.

Third, if you do get the problem wrong, it will be easier to locate errors when the instructor reviews the test. This prevents you from making the same mistakes on your next math test.

Fourth, showing your work shows your instructor that you put in the effort to solve a problem and did not just guess or leave it blank. Many college professors are more generous than you might think. If they can tell that you put in the work, they may give you enough points to pass a test you might have otherwise failed. The important thing is to put in the effort. Even if you occasionally come up short, you'll never embarrass yourself if put in the right amount of work.

Activity 7.1 Assessing Your Test-Taking Methods

By reviewing how you took your last math test we can determine what changes you can make to improve your test-taking skills. In the space below, list your typical test-taking steps and explain them. Some students may have less than six steps and others may have more than six steps. If you have more than six steps, add them in the blank space at the bottom of this page.

Step One:

Step Two:

Step Three:

Step Four:

Step Five:

Step Six:

How to Become a Better Computer-Based Test Taker

Taking a computer-based math test differs in many ways from taking tests in traditional classrooms. Whether a part of an online math course or an Emporium classroom, computer-based tests usually measure two areas: the content of the material and how well you take a test. The second measure is more important for online tests than classroom tests and requires different test-taking skills.

First of all, depending on the rules and setup of the software you are using, you may have to vary your plan of attack. Different tests allow you different amounts of time to complete problems. Also, some tests allow you to revisit previous questions and some don't. I've worked with many students who have lost test points because a test was turned off after 55 minutes, and they were still working on problems. I've seen others lose points because the software prevented them from going back and checking their answers.

Before deciding which of the Ten Steps to Computer Test-Taking you want to use, ask your instructor these questions: Where can I find a hard copy of sample practice problems or a practice test? How much time do I have for the quiz or test? How many problems are on the quiz or test? Will the quiz or test automatically terminate after a certain period of time? Can I have extra test time if I did not finish a problem? Can I go back and review the problems that I do not finish? Can I go back and review the problems that I finish? Can I turn in my scratch paper for partial credit? Once the tests starts, can I use a blank sheet of paper for my memory data dump (mental cheat sheet)? After the test is graded, can I go over the test to review my mistakes?

Each instructor in a distance-learning class or Emporium model class may set up the testing process differently. Emporium tests may or may not occur during class time. It is important that you understand how this process works during the first week. It is possible that you may not be able to take the tests in the way that they are set up. Math tests in distance-learning classes have been particularly problematic. Review the syllabus to see how and where you will be tested.

In most cases, you will take practice tests and quizzes online, but the major tests will probably be in a secure test center, usually on a college campus. In some cases you may have to go to the local community college or university and pay to take the test in their testing centers. Make sure you are able to do this if it is the way the test administration is set up. If you can't, talk to the instructor immediately during the first week of classes. This way, if there are no alternative ways to take tests, you are still allowed to drop or change sections of the math class.

10 Steps to Taking a Computer-Based Math Test

Before taking your test, practice the steps printed below.

Step One—*When you get your test, use the blank sheet of paper you showed to your instructor, and write down the information from your mental cheat sheet.* Your mental cheat sheet has now turned into a list, and writing down this information is not cheating. Do not wait until you get stuck on a problem before writting down the memory data dump. By that time, you may have forgotten the information you tried to memorize. Write down facts, figures and formulas from your mental cheat sheet or other information you might not remember during the test.

Step Two— *Develop a test progress schedule.* When you begin setting up your test schedule, determine how much time you have for each question. You should already know how many questions are on the quiz or test. Then, figure out how many minutes you have to complete each problem. For example, if you have 25 test problems to complete in 60 minutes, then you have about two minutes per problem. Decide how many problems should be completed halfway through the test. You should have more than half the problems completed by that time.

Step Three— *Start by answering the problems in proper order, unless your program allows you to skip and return to problems.* Solve the problems quickly and accurately. Then, review the answers to see if they make sense. On most computer math tests, the easier problems are near the beginning; you need to answer them efficiently and quickly. This will give you more time for the harder problems and time to review the whole test (if the software allows). If you can choose the order you answer questions, always do the easiest problems first.

Step Four— *Start working through the rest of the test as fast as you can while remaining accurate and thoughtful.* Answers should be reasonable. Make sure you are using scratch paper to work problems. Clearly write down each step because if you are allowed to turn in your scratch paper, you may get partial credit even if you end up missing the problem. Also, if you cannot turn in the scratch paper with the test, you can still use the scratch paper when reviewing. This allows you to discover your mistakes and correct them.

Step Five— *Second memory data dump.* The second data dump is for writing down material that was forgotten or jarred while working on other problems. If you are working on a problem and remember something you forgot to write on your memory data dump, write it down now. If you wait to finish the problem, you may forget it.

Step Six— *Try difficult problems.* As you are working through your test, you may encounter a difficult problem. These are the problems that you might remember something about, but you don't know enough to solve them. You might remember the first step, but then draw a blank on what to do next. It is important to begin the problem, even if you know that you cannot complete it. By starting the problem and writing down what you know, you are warming up your brain, which may lead to the solution. Take time on these problems. If you get stuck and are able to return to unsolved problems, write down the problem number and move on. **If your software does not allow you to review problems, then guess and move on to Step 10.**

Step Seven— *Review skipped problems.* While you are working problems, certain information may trigger an "Ah ha!"response. The "Ah ha!" response is when your brain suddenly remembers how to complete a problem. A memory from your homework or a lecture sometimes pops back into your head in the middle of a test. When this happens, immediately go back to the skipped problem and solve it. If you did not have an "Ah ha!" response while finishing other problems, go back and review the skipped problems and try to solve them anyway.

Step Eight— *Guess at the remaining problems.* If you cannot write down the first step, rewrite the problem. Also, remember that the way you learned how to solve a problem was by writing—not just looking at the problem. Sometimes, rewriting a problem jars your memory enough to complete the first step or the entire problem. Also, review your memory data dump to see if any of that information leads to writing the first step.

If you are taking a multiple choice test, take the answers and put them back into the equation to see which one is correct. Do the opposite of the function to see if you can work backwards to get the correct answer. For example, if you are asked to simplify $5(x + 2)$, then take the answer and do the opposite of the function to see if your answer is correct. The answer is $5x + 10$. The opposite of the function involves factoring $5x + 10$, which is $5(x + 2)$. This is the correct answer because it matches the problem you started with. Based on the time you are given, you may only solve a few problems using this strategy. However, even if you only solve a few problems you will still gain test points. If you still don't the answer to a problem, choose the middle range of potential answers. Often times, the correct answer is neither the highest nor lowest number listed in multiple choice fields. Whatever you do, always remember to select an answer for every problem.

Step Nine— *Review the test.* Review each problem by substituting the answer back into the equation or doing the opposite function required to answer the question. If you cannot check the problem in these ways, rework the problem on a separate sheet of paper and compare the answers. Look for careless errors or other errors you may have made. Students usually lose two to five test points on errors that could have been caught in review. Do not talk yourself out of an answer just because it was easy to solve.

Step Ten— *Use all of your alloted test time.* Do not close the test unless you have reviewed each problem twice, or, if you can't review problems, until you have finished all requirements. When working on the last problems, don't rush. This is an excellent time to make sure you get problems correct.

Even though we encourage students to use all of their test time, most don't. After talking to hundreds of these students, I have discovered two different themes for leaving the classroom early. First, test anxiety gets so overwhelming that they cannot continue the test. The only way to lessen this anxiety is to finish the test before time is up—even though they may get a lower grade. Second, in the Emporium model classroom, or in a testing center, students sometimes rush through a test to avoid being the last to finish. They feel as though other students are watching and that the students who finish last are "dumb and stupid." These students also believe that students who hand their tests in first make "A's" and "B's,"and those students who finish their tests last make "D's" and "F's." If you are one of these students, you don't need to care about what other students think about you (it's usually wrong anyway). YOU need to fight the urge to finish early. Use all your alloted test time.

Activity 7.2 Assessing Your Test-Review Methods

Before you can learn how to turn your old tests into handy study guides, you need to determine what you are currently doing with your tests and how well this method is working for you. Below, put a check by what you tend to do with your graded tests (you can mark more than one):

_____ Throw the test away

_____ Look at the grade and put the test in your book

_____ Review the test without working out missed problems

_____ Review the test and work out missed problems

_____ Compare the problems you missed with another student's problems

_____ Work problems that are similar to the ones that you missed

_____ Go over the missed problems with your tutor

_____ Go over the missed problems with your instructor

_____ Use the test review to develop new test-taking strategies

Other:

The Six Types of Test-Taking Errors

When students get their tests back they usually don't look at them. Some throw them away or stick them in their books and forget about them. In most cases it does not matter if the students make a good grade on the test or a poor grade. They do not like reviewing the tests. Some instructors make the students review the tests by going over them in class, but rarely do the students analyze the tests on their own. This leads to repeating mistakes.

As a comparison, imagine what would happen if your doctor gave you a blood test to see why you were sick and you decided to just throw away the results and just pick some medicine off the shelf to see if it would cure you. The doctor would be sued for malpractice, and you may pick the wrong medicine and die. Perhaps the analogy exaggerates, but it makes the point. The real question is, "What do you do with your returned tests?"

Analyzing your math test will not kill you, but not doing it may kill your grades. Many students make the same mistakes over and over again, which causes them to fail too many tests. Students who analyze their tests are at a great advantage to determine how to improve on their next tests. These students find out the types of mistakes they made and now can change them. They develop better test-taking strategies for the next test and in most cases make higher scores. To improve future test scores, you must conduct a test analysis of previous tests.

In analyzing your tests, look for the following kinds of errors:

1. Misread-direction errors
2. Careless errors
3. Concept errors
4. Application errors
5. Test-taking errors
6. Study errors

Look at your last test to determine how many points you lost in the six types of test-taking errors. Analyze any previous tests to determine if there is a pattern of test errors. The type of errors is very important to determine how you can improve your studying and test-taking. For more information, visit the sample test printed at the back of this chapter.

Misread-Direction Errors

Now look at your last test. See how many points you lost to misread-direction errors. Misread-direction errors occur when you skip or misunderstand directions and do the problem incorrectly. If you do not understand if a problem you missed was a result of misread directions, ask the instructor for clarification.

Example 1: You have this type of problem to solve: (x + 1) (x + 1). Some students will try to solve for x, but the problem only calls for multiplication. You would solve for x only if you have an equation such as (x+1)(x+1) = 0.

Example 2: Another common mistake is not reading the directions before doing several word problems or statistical problems. All too often, when a test is returned, you find only three out of the five problems had to be completed. Even if you did get all five of them correct, it costs you valuable time, which could have been used obtaining additional test points.

To avoid misread-direction errors, carefully read and interpret all the directions. Look for anything that is unusual, or if the directions have two parts. If you do not understand the directions, ask the instructor for clarification. If you feel uneasy about asking the instructor for interpretation of the question, remember the instructor in most cases does not want to test you on the interpretation of the question but how you answer it. Also, you don't want to make the mistake of assuming that the instructor will not interpret the question. Let the instructor make the decision to interpret the question, not you.

Careless Errors

Careless errors are mistakes made because students lose complete focus on the question. If a student is nervous or in a hurry, little mistakes like switching signs mess up the entire problem. Sometimes, simple problems turn into nearly impossible problems if a student makes a careless error. This can really waste precious test time. Careless errors can be caught automatically when reviewing the test. Both good and poor math students make careless errors. Such errors can cost a student the difference of a letter grade on a test. Many students want all their errors to be careless errors when they analyze their tests because it is easier to admit carelessness than admitting that they just didn't know the material. In such cases, I ask the student to solve the problem immediately while I watch. If the student can solve the problem or point out his/her mistake in a few seconds, it is a careless error. If the student cannot solve the problem immediately, it is not a careless error and is probably a concept error.

Careless Error Examples

1. Dropping the sign: -3(2x) = 6x, instead of -6x, which is the correct answer.
2. Not simplifying your answer: Leaving (3x -12)/3 as your answer instead of simplifying it to x - 4.
3. Adding fractions: 5/16 + 7/16 = 12/16 instead of 3/4.
4. Word problems: X = 15 instead of the "student had 15 tickets."

When working with students who make careless errors, I ask them two questions: First, "How many points did you lose due to careless errors?" Then I follow with, "How much time was left in the class period after you handed in your test?" Students who lose test points to careless errors are giving away points if they hand in their test papers before the test period ends.To reduce careless errors, you must realize the types of careless errors made and recognize them when reviewing your test. If you cannot solve the missed problem immediately, it is not a careless error. If your major error is not simplifying the answer, review each answer as if it were a new problem and try to reduce it.

Concept Errors

Concept errors are mistakes made when you do not understand the properties or principles required to work the problem. Concept errors, if not corrected, will follow you from test to test, causing you to lose a lot of points.

Some common concept errors are not knowing:

- (-)(-)x = x, not -x
- -1(2) > x(-1) implies 2<x, not 2>x
- 5/0 is undefined, not "0"
- (a+x)/x is not simplified to "a"
- The order of operations
- 2 + 3 x 5 = 17, not 25
- 1/2 + 1/3 = 5/6, not 2/5

Concept errors must be corrected to improve your next math test score. Students who have numerous concept errors will fail the next test since each chapter builds on the previous ones. Just going back to rework the concept error problems is not good enough. You must go back to the textbook or notes and learn why you missed those types of problems, not just the one problem itself.

The best way to learn how to work those types of problems is to set up a concept-problem error page in the back of your notebook. Label the first page "Test One Concept errors." Write down all your concept errors and how to solve them. Then, work five more problems, which use the same concept. Now, in your own words, write the reasons that you can solve these problems.

If you cannot write the concept in your own words, you do not understand it. Get assistance from your instructor if you need help finding similar problems using the same concept or cannot understand the concept. Do this for every test. For example, when preparing for your second test, review the concept errors on the first test. Continue the process all the way into the final. This is a proven system to improve your test scores.

Application Errors

Application errors occur when you know the concept but cannot apply it to the problem. Application errors usually are found in word problems, deducing formulas (such as the quadratic equation) and graphing. Even some better students become frustrated with application errors; they understand the material but cannot apply it to the problem.

To reduce application errors, you must predict the type of application problems that will be on the test, then think through and practice solving those types of problems using the concepts.

> Example: If you must derive the quadratic formula, you should practice doing that backward and forward while telling yourself the concept used to move from one step to the next.

Application errors are common with word problems. When solving word problems, look for the key phrases displayed in the "Translating English Words into Algebraic Expressions" figure to help you set up the problem. After completing the word problem, reread the question to make sure you have applied the answer to the intended question.

Application errors can be avoided with appropriate practice and insight. However, if all else fails, memorize a word problem or a graphing problem and use it as part of your memory data dump. Write it down on your test and use it as an example to solve the problems on the test.

Test-Taking Errors

Test-taking errors apply to the specific way you take tests. Some students consistently make the same types of test-taking errors. Through recognition, these bad test-taking habits can be replaced by good test-taking habits. The result will be higher test scores. The list that follows includes the test-taking errors, which can cause you to lose many points on an exam.

Error 1 — *Missing more questions in the first third, second third or last third part of a test.* Divide your test in thirds. For example, if your test had 20 problems, look at the number of points missed from problems 1 through 6. How many points were missed from problems 7 through 13? How many points were missed from problems 14 to 20? Missing more questions in the first third of a test could be caused by carelessness when doing easy problems or from test anxiety. Missing questions in the last part of the test could be due to the fact that the last problems are more difficult than the earlier questions or due to increasing your test speed to finish the test.

If you consistently miss more questions in a certain part of the test, use your remaining test time to review that section of the test first. This means you may review the last part of your test first.

Error 2 — *Not completing a problem to its last step.* If you have this bad habit, review the last step of the test problem first before doing an in-depth test review.

Error 3 — *Changing correct answers to incorrect answers.* Look for erased answers on your test. I have seen students erase the correct way to solve the problem, and thus the correct answer, and then change it to an incorrect answer. I have asked these students why they erased the correct answer. In almost every case the student said that the answer was too easy to solve, so they thought it was wrong. These students need to build their test-taking confidence and know how to check their answers. If this is you, make sure you learn how to check your answers and have confidence in them.

On multiple choice tests, most students indicate that they are bad answer changers. That is because they only remember changing answers from right to wrong. They usually do not remember the times they change answers from wrong to right.

Now, let's see if you are a good, bad or neutral answer changer.

On multiple choice problems, you can change your answer three ways: from right to wrong, wrong to right and wrong to wrong. Look at the eraser marks on your answer sheet that represent the number of answers that you changed. You may want to do this on several of your tests. Count the number of changes that represent the three categories and change them into percents.

If you have a pattern of changing over 50 percent of your answers from wrong to right then keep changing your answers. If over 50 percent of your answers are changed from right to wrong then stop changing answers. If the three categories of changing answers are about the same, then do what makes you feel the best because it is not affecting your grade.

If you don't have previous tests to measure the effectiveness of your answer changing, then wait to review several tests before you make your decision. If you are indeed a bad answer changer, then on test day use a pencil without an eraser to mark your multiple choice tests. This should stop you from changing your answers.

Error 4 — *Getting stuck on one problem and spending too much time on it.* You need to set a time limit on each problem before moving to the next problem. Working too long on a problem without success will increase your test anxiety and waste valuable time that could be used in solving other problems or in reviewing your test. For example, if your test has twenty problems that must be completed in one hour, how long should you spend on each problem? You should spend the maximum of three minutes on each problem. Of course this depends on the complexity of the problems. Spending more than three minutes, unless it is the last problem you are working on, can cost you test points.

Error 5 — *Rushing through the easiest part of the test and making careless errors.* This happens even to the best students. If you have the bad habit of getting more points taken off for the easy problems than for the hard problems, first review the easy problems and then the hard ones.

Error 6 — *Incorrectly copying an answer from your scratch paper to your test.* To avoid these kinds of errors, systematically compare your last problem step on scratch paper with the answer written on the test. In addition, always hand in your scratch work with your test.

Error 7— *Leaving answers blank.* If you look at a problem and cannot figure out how to solve it, do not leave it blank. Write down some information about the problem, rewrite the problem or try to do at least the first step.

Error 8— *Answering only the first step in a two-step problem.* These students get so excited when answering the first step of the problem that they forget about the second step. This is especially true on two-step word problems. To correct this test-taking error, write "two" in the margin of the problem. That will remind you that there are two steps or two answers to this problem.

Error 9— *Not understanding all of the functions of your calculator.* Some students barely learn how to use the critical calculator functions. Then they forget or have to relearn how to use their calculator during the test, which costs test points and time. Do not wait to learn how to use your calculator on the test. Over-learn the use of your calculator *before* the test.

Error 10—*Completing the test early without checking all of your answers.* Do not worry about the first person who finishes the test and leaves. Many students get nervous when other students start to leave early. This can lead to test anxiety, mental blocks and loss of recall. According to research, the first students to finish a test do not always get the best grades. It sometimes is the exact opposite. Ignore those students who leave and use the full time allowed.

Error 11—*Trying to complete problems in your head.* This is especially common among distance learners and students in Emporium model courses. These student get into this habit while doing online homework. Doing problems in your head causes you to lose points. Make sure to write down your steps so that you are able to check your work.

Error 12—*When taking a timed test on a computer, many students do not use the last few minutes effectively.* If students cannot review their test answers, they do not slow down and check all the last questions to make sure the problems are correct. If they can check their answers, then they do not go back and do it because they think it is a waste of time. Both types of students are wasting time and losing test points.

Study Errors

Study errors, the last type of mistake to look for in a test analysis, occur when you study the wrong type of material or do not spend enough time on pertinent material. Review your test to find out if you missed problems because you did not practice that type of problem or because you did practice it but forgot how to do it during the test. Study errors will take some time to track down, but correcting study errors will help you on future tests.

Most students, after analyzing one or several tests, will recognize at least one major, common test-taking error. Understanding the effects of this test-taking error should change your study techniques or test-taking strategy.

> Example: If there are seven minutes left in the test, should you review for careless errors or try to answer those two problems you could not totally solve? This is a trick question. The real question is, "Do you miss more points due to careless errors or concept errors, or are the missed points about even?" The answer to this question should determine how you will spend the last minutes of the test. If you missed more points due to careless errors or missed about the same number of points due to careless/concept errors, review for careless errors.

How to Prepare for a Final Exam

The first day of class is when you start preparing for the final exam. Look at the syllabus or ask the instructor if the final exam is cumulative. A cumulative exam covers everything from the first chapter to the last chapter. Most math final exams are cumulative. The second question you should ask is if the final exam is a departmental exam or if your instructor makes it up. In most cases, departmental exams are more difficult and need a little different preparation. If you have a departmental final, you need to ask for last year's test and ask other students what their instructors say will be on the test.

The third question is, how much will the final exam count? Does it carry the same weight as a regular test or, as in some cases, will it count for a third of your grade? If the latter is true, the final exam will usually make a letter grade difference on your final course grade. The final exam could also determine if you pass or fail the course. Knowing this information before the final exam will help you prepare. You also need to calculate the numerical grade you need to make on the final to pass your course with the grade you want. Knowing this before the exam will help decide how much study time you need.

The fourth question is, do students have to pass the final exam to pass the course. Some math departments set up final exams as an exit exam to the next course. They know that the final exam is an accumulation of all the course knowledge and predicts success in the next course. Knowing all this information before taking the finial will help you prepare.

Preparing for the final exam is similar to preparing for each chapter test. You must create a pretest to discover what you have forgotten. You can use questions from the textbook chapter tests or questions from your study group. Review the concept errors that you recorded in the back of your notebook labeled "Test One," "Test Two," etc. Review your problem log for questions you consistently miss. Even review material that you knew for the first and second test but which was not used on any other tests. Students forget how to work some of these problems.

If you do not have a concept error page for each chapter and did not keep a problems log you need to develop a pre-test before taking the final exam. If you are individually preparing for the final, then copy each chapter test and do every fourth problem to see what errors you may make. However, it is better to have a study group where each of the four members brings in ten problems with the answers worked out on a separate page. Then each group member can take the 30-question test to find out what they need to study. You can refer to the answers to help you solve the problems that you miss. For an example of how to analyze a previous test, review the boxes printed on the next page.

Math Test for Prealgebra

The answers are in boldface. The correct answers to missed questions are shaded. Identify the type of error based on the Six Types of Test-Taking Errors. The student's test score is 70. The answer key is on the next page.

1. Write in words: 32.685

   Thirty-two and six hundred eighty-five thousanths

2. Write as a fraction and simplify: 0.078

   $\frac{78}{1000}$ $\frac{39}{500}$ -2

3. Round to the nearest hundredth: 64.8653

   64.865 **64.87** -4

4. Combine like terms: $6.78x - 3.21 + 7.23x - 6.19$

   $= 6.78x + 7.23x + (-3.21) + (-6.19)$
   $= 14.01x - 9.4$

5. Divide and round to the nearest hundredth: $68.1357 \div 2.1$

   32.4454 → 32.45

6. Write as a decimal: $\frac{5}{16}$

   0.3125

7. Insert < or > to make a true statement.

   $\frac{3}{8} < \frac{6}{13}$

8. Solve: $\frac{3}{x} = \frac{9}{12}$ -2

   $9x = 3(12)$

   $\frac{x}{9} = \frac{36}{9}$

   $x = 5$ $x = 4$

9. What number is 35% of 60? -4

   2100 **21.00**

10. 20.8 is 40% of what number? -8

    52

11. 567 is what percent of 756? -8

    $\frac{756}{567} = \frac{9}{100}$

    $= 133.3\%$ 75%

12. Multiply: -2

    $(-6.03)(-2.31) = 13.9$ 13.9293

Answer Key for Prealgebra Test

1. Correct
2. *Misread-directions error*—forgot to simplify by reducing the fraction.
3. *Concept error*—did not know that hundredths is two places to the right of the decimal.
4. Correct
5. Correct
6. Correct
7. Correct
8. *Careless error*—divided incorrectly in the last step.
9. *Test-taking error*—did not follow step 5 in test-taking steps: reviewing answers to see if they make sense. The number that equals 35% of 60 can't be larger than 60.
10. *Test-taking error*—did not follow steps 7 and 8 in test-taking steps: don't leave an answer blank.
11. *Application error*—solved the equation correctly but the equation setup was wrong.
12. *Concept error*—did not know that when you multiply with one number in the hundredths, the answer must include the hundredths column.

Chapter Seven: Checklist for Knowledge and Behaviors

Put a check after understanding the knowledge statements and fill in the behavioral activities.

Chapter Knowledge

1. I understand that just completing homework and attending class does not guarantee that I will pass my math tests. ____
2. I understand that I should not cram for math tests. _________
3. I know that improving my math test-taking skills begins with completing a practice math text several days before the actual test. This pretest can help me locate math areas that I need to study. _______
4. I can improve my math weakness by reviewing the homework in that area or obtaining help from my instructor or tutor. ________
5. I know to follow the 10 Steps to Taking a Math Test to obtain the greatest number of points in the least amount of time. ________
6. I know that students who turn their test in first or last make "A's" and "B's". ____
7. I know that without conducting a test analysis, I will probably continue to make the same old test errors and lose valuable test points. ______
8. I need to decide before each test if I am going to spend the last few minutes of each test checking for careless errors or finishing problems you left in completed. _______
9. I know to review my Concept Errors from each previous test(s) before the next test. _____
10. I know when preparing for the final exam, make sure to take a pretest developed by either the instructor, your-self or study group. This pretest must be taken several days ahead of time under the same timed conditions as the final exam. _____

Behavioral Activities

1. I start studying for the test three days ahead of time. ________
2. I only review already learned material the night before the tests________
3. I have developed my own memory data dump for my next test. ____
4. During the test I will stay in the math classroom until each answer is checked or until time is up. _____
5. I will use my chosen short -term test anxiety reduction technique during the test. _____
6. I developed my own steps to taking a math test (at least 5) . ___
7. I recorded Concept Error problems from each test in the back of my notebook and review them before each test. ______
8. When I get my tests back I met with the tutor to review the missed problems. ____
9. I keep all the tests that are returned and review them before the next test.____
10. I will form a study group to prepare for the final exam. ________

Name: ______________________________ Date: ______________

Assignment for Chapter 7

Answer the following questions on a separate piece of paper.

1. List four reasons for why you can understand how to do the math problems in class but can still miss them on the test.

2. List four reasons how you can correctly complete all the homework assignments and still not score highly on math tests.

3. List two ways you can use a study group to prepare for a math test.

4. List and discuss two general pretest rules that can help you become more successful in taking your math test.

5. List three reasons you make a plan before taking a math test.

6. List the Ten Steps for Better Test-Taking (either online or in-class).

7. List the Six Types of Test-Taking errors:

8. Complete a test analysis on your last test. List two types of errors you made and how you can correct them.

9. What are the answers to the three questions you need to ask your instructor about the final exam?

10. How should you run a study group to prepare for tests and the final exam?

A

APPENDIX

Learning Modality Inventory for Math Students

The following survey can help you discover how you best learn math. Answer the questions based on what you are like. There are no right or wrong answers. The more you answer truthfully, the more you will be able to use the results to improve studying math. "1" means the statement is hardly like you. "4" means the statement is really like you. Then, if you think the statement is somewhere in between, decide if it is a "2" or a "3."

Questions	Least like me		Most like me	
1. Reading a math problem out loud helps me learn better when I am studying.	1	2	3	4
2. I learn math better if I can talk about it.	1	2	3	4
3. I select certain problems and memorize what they look like so I can use them to help me remember on a math test.	1	2	3	4
4. Making things with my hands helps me learn better.	1	2	3	4
5. Drawing a picture of the word problem helps me understand how to do it on a test.	1	2	3	4
6. Math makes more sense when I see it worked out on the board.	1	2	3	4
7. Moving around while studying helps me concentrate and learn more.	1	2	3	4
8. I understand written instructions better than ones told to me.	1	2	3	4
9. I memorize what a problem looks like so I can remember it better on a test or quiz.	1	2	3	4

Questions	Least like me		Most like me	
10. I repeat steps to a problem out loud or to myself in order to remember what I am supposed to do.	1	2	3	4
11. Watching someone complete a math problem helps me understand more than listening to someone tell me how to do it.	1	2	3	4
12. Talking about a math problem while learning in class helps me understand it better.	1	2	3	4
13. I learn math better when I watch someone do it.	1	2	3	4
14. When I take a test, I read the problems to myself softly.	1	2	3	4
15. When I solve a math problem on a test, I picture my notes in my head to help me remember how to solve it.	1	2	3	4
16. I enjoy making things with my hands for a hobby.	1	2	3	4
17. Math makes more sense when someone talks about it while doing it on the board rather than just doing it on the board.	1	2	3	4
18. Explaining a math problem to someone else helps me learn better when I am studying.	1	2	3	4
19. Looking at a picture from my notes or math book helps me understand a math problem.	1	2	3	4
20. Making study aids with my hands helps me learn better.	1	2	3	4
21. I understand instructions better when someone tells me what they are.	1	2	3	4
22. I memorize sentences or words I can say to myself to help me remember how to do problems on a test.	1	2	3	4
23. Pictures and charts help me see how all the parts of a word problem work together.	1	2	3	4
24. I enjoy putting things together.	1	2	3	4
25. When I solve a problem on a math test, I talk my way through it in my head or softly to myself.	1	2	3	4

Scoring Your Results

Step One: Fill in each answer score in the appropriate question number. Add the column totals. Divide Column Totals A and B by 2. Those numbers will be your final column totals. Leave Column C total as is.

Column A	Column B	Column C
1. ________	3. ________	4. ________
2. ________	5. ________	7. ________
10. ________	6. ________	16. ________
12. ________	8. ________	20. ________
14. ________	9. ________	24. ________
17. ________	11. ________	
18. ________	13. ________	
21. ________	15. ________	
22. ________	19. ________	
25. ________	23. ________	
A total ______/2 =_____ Column	B total ______/2 =_____ Column Total	C Total ________ (Do not divide).

Step Two: Fill in the number of squares to represent each column total. Any total greater than 12 indicates that modality style as a strength when you learn math. You can be strong in more than one modality. If none of the totals equal 12 squares, your highest score is your strongest modality. If you have a tie, pick the first one that comes to mind as your strongest.

	Least Like Me																			Most Like Me
Modality	1				5					10					15					20
A = Auditory																				
B = Visual																				
C = Kinesthetic																				

B
APPENDIX
My Math Success Plan

Semester:

A: Student Information
Name:
Year in College:

B: Course

C: Learning Information

D: Semester Goals

1.

2.

3.

4.

5.

6.

E: Math Study Strategies

F: Motivation Strategies

C

APPENDIX

Apps and Websites for Academic Support

The apps and websites below are resources to improving learning. Some of the apps are for recording information and other apps and Web sites are for learning math and checking homework problems. Review these resources to discover which one are best for you.

Recording Apps

1. Smart Voice Recorder – Recorder and label recording
2. Evernote – Type in notes, take photo of notes, record notes, attach a file, hand write and sync with computer.
3. Voice Recorder - can record information on your smart phone.
4. Photos - Can show photos of problems from the board or any written material to your tutor or instructor.
5. Video Recorded - Can video record lectures or tutoring.

Learning Math and Checking Answers

1. My Script Calculator – can write in arithmetic problems and solve them.
2. Algebra Tutor- solves arithmetic and algebra problems and has practice problems also has a learn function.
3. Algeo – multiply, Squares, sin, cos graphing solutions, and f(x).
4. Photomath – take pictures of equations and solves them on your phone.
5. Web Math Algebra - https://tinyurl.com/c4bo7gf - This is a free site to help solve algebra problem. Also go to the bottom of the page to see similar apps for high school and college.
6. Wolfram Algebra app - http://products.wolframalpha.com/courseassistants/algebra.html - Algebra course assistant for high school and college and the App cost of $1.99.
7. Wolfram Alpha - http://www.wolframalpha.com/ - This site does all math levels.
8. Purple Math - www.Purplemant.com - This is one of the oldest sites for all levels of algebra.
9. Khan academy - www.khanaccademy.com - This site has videos on solving all levels of math problems and you can select the concept to be learned.
10. You Tube - www.YouTube.com - This site has videos on doing all levels of math problems. Type in the level of math such as Intermediate Algebra or the concept such as factoring trinomials.

Bibliography

Ashcraft, M & Kirk, E. (2001). "The relationships among working memory, math anxiety, and performance." *Journal of Experimental Psychology*: General, 130(2), 224-237.

Bandura, A. (1982). "The psychology of chance encounters and life paths." *American Psychologist*, 37, 747-755.

Bloom, B. (1976). *Human Characteristics and School Learning*. New York: McGraw-Hill Book Company. (p. 35)

Brown, J. & Copper, R. (1978). *Learning Styles Inventory*. Kirkwood, MO: Educational Activities Software. (p. 46)

Brunye, T., Mahoney, C., Giles, G., Rapp, D. Taylor, H., & Kanarek, R. (2013). "Learning to relax: Evaluating four brief interventions for overcoming the negative emotions accompanying math anxiety." *Learning and Individual Differences*, 27, 1-7.

Butler, P. (1981). *Talking to Yourself. Learning the Language of Self-Report*. San Francisco: Harper & Row. (p. 219)

Farr, M & Ludden (2013) *Best Jobs for the 21st Century*. Indianapolis, IN: JIST Works pp 84-87 (p. 25)

Gray, S. (1991). "Ideas in Practice: Metacognition and Mathematical Problem Solving," *Journal of Developmental Education*, 14 (3), 24 - 30.

Nolting, K. (2008). *Learning Modality Inventory for Math Students*. Bradenton, FL: Academic Success Press, Inc. (p. 46)

Nolting, P. (1987). *How to Reduce Test Anxiety, a CD*. Bradenton, FL: Academic Success Press, Inc. (p.40)

Nolting, P. D. (2014). *My Math Success Plan, a Student Workbook*. Bradenton, FL: Academic Success Press, Inc. (p. 40)

Richardson. F, and Suinn, R. (1973). "A comparison of traditional systematic desensitization , accelerated mass desensitization, and mathematics anxiety." Behavior Therapy . No. 4, pp. 212-218. (p. 57)

Schunk, D. H. (1991). "Self-efficacy and academic motivation," *Educational Psychologist*, 26, 207-231.

Spielberger, C. (1980). *Test Attitude Inventory*. Redwood, California: Mind Garden.

Tobias, S (1978). "Who's Afraid of Math and Why?" *Atlantic Monthly*, September, pp. 63-65. (p. 57)

Zientek, L. , Ozel, Z. , Fong, C., Griffin, M. (2013). "Student Success in developmental mathematics courses," *Community College Journal of Research and Practice*, 17, 990 - 1010.

About the Author

Copyright 2014 Randy Kennedy

Over the past 30 years, Dr. Paul and Kimberly Nolting have helped thousands of students improve their math learning skills and obtain better grades. They have consulted with over 100 colleges and universities. Dr. Nolting is a national expert in assessing math learning problems, developing effective student learning strategies, assessing institutional variables that affect math success, tutor training and math study skills. He is also an expert in how to help students with disabilities and Wounded Warriors become successful in math. He has been a undergraduate instructor, a graduate instructor, a learning specialist, an institutional test administrator, Student Support Services Director, Title III director, and a math lab and disability coordinator.

Dr. Nolting holds a Ph.D. in Education in Curriculum and Instruction from the University of South Florida. His dissertation was on using math study skills to improve academic achievement. His expertise ranges from designing math study skills curriculum to redesigning math courses. He is also an expert in disabilities as they relate to mathematics learning and how to set up effective accommodations. Since his dissertation in 1986 on improving math success for developmental math students with math study skills intervention, he has continued to conduct mathematics research, consult, serve as an interview subject for journal articles, write journal articles, and conduct national training.

Dr. Nolting has conducted numerous national workshops on math learning and disability issues at the American Math Association of Two Year College (AMATYC), American Mathematics Association (MAA), National Developmental Education Association (NADE), and the Association on Higher Education and Disabilities (AHEAD). He has been the keynote speaker at many state conferences. He was a consultant for the American College Test (ACT) and the Texas Higher Education Coordinating Board.

Dr. Nolting has been widely acclaimed for his ability to communicate with faculty and students on the subject of improving math learning. He now helps colleges and universities redesign math courses to meet new curriculum requirements, and also assists institutions improve retention rates. He also helps devolop Quality Enhancement Plans (QEPs), one of which one won QEP of the year, state wide mandates and improving retention. His team is currently working on a QEP that will redesign a college's developmental math curriculum to better integrate math study skills, manipulatives and tutoring. The pilot course has a 71% pass rate.

Dr. Nolting is also a nationally recognized consultant and trainer of math study skills, math faculty training, tutor/learning assistance training and improving classroom success. He was interviewed by Dr. Hunter Boylan as a national expert in the Journal of Developmental Education. The article was titled, "Improving Success In Developmental Mathematics: An Interview with Paul Nolting." He was also interviewed by U.S. News and World Report about developmental math and online math success. Finally, he coordinated two National Math Summits: the first as a pre-conference to the 2013 AMATYC national conference, and the second as a pre-conference to the 2016 NADE national conference. The summits were a response to a nationwide need for mathematics course redesigns and assessments. The summits featured panels of national experts along with workshops to help college faculty and administrators develop their own, "Math Success Plan for College Innovation." Between those summits he also coordinated math panels at the NADE, AMAYC and MAA conferences.

He is also well-regarded for his ability to teach faculty and tutors about learning styles. This process includes assisting administrators as they develop affective math and study.

Over the last thirty years, Dr. Nolting has consulted with over 100 colleges, universities and high school campuses, including the University of Massachusetts, Florida State University, New Jersey City University, Rucker University, University of Colorado-Bolder, Texas Tech University, Black Hills State University, Tennessee Tech University, Austin Pea University, Clemson University, Colorado State University and the University of Connecticut, and Moorehead State.

Other colleges with which he has consulted include San Antonio College, St. Louis Community College, J. Sargeant Reynolds College, Montgomery College, Broward College, Miami-Dade College, Northeast State Technical Community College, Landmark College, Denver Community College, Valencia College, Mesa Community College, Glendale Community College, and Austin Community College.

He also has written seven other books, three DVDs, two CDs and authored three websites. His latest two books are *My Math Success Plan* and *Mathematics and Disability Handbook*. The latter is a workbook for wounded warriors and students with disabilities, which combines math study skills and educational information on how disabilities affect math learning. The Mathematics and Disability Handbook serves as a reference text for math instructors and disability/veteran coordinators.

Rochelle Beatty

The Quick Start section in Chapter One is designed to help students immediately start using math study skills in lecture, emporium model and modular courses. In the past, students have had to delay using certain math study skills until they were discussed in each chapter. Now students can begin using math study skills in the first week of their course and continue to build these skills with each chapter. I would like to thank Dr. Rochelle Beatty, Assistant Developmental Mathematics Professor at Kansas City Kansas Community College (KCKCC) for her contribution to this section. Dr. Beatty is one of the coordinators for KCKCC's emporium model mathematics courses, where computer-assisted instruction is the design for most course sections. She also consults with colleges on other math designs such as modular. She has been working with faculty and students for over 20 years on improving math success. Her suggestions provided in this section are practical and proven.

Denise Turcotte

The new Check List for Knowledge and Behaviors was designed for students to check off that they understand certain facts about math study skills and have completed certain study skills behaviors associated with math success. If they do not understand the math study skills knowledge then they need to reread that section. The behaviors can be checked off and turned in to the instructor with any additional required work. Completing these sections allows students to know and practice good math study skills, test anxiety reduction and test-taking skills. I would like to thank Ms. Denise Turcotte who is an Advisor III and adjunct math instructor at State College of Florida for suggesting some of the knowledge and behaviors statement and other ideas in the book. She also has been a math learning specialist and has helped thousands of students improve their math success. Her suggestions have helped students pass math.

Index

T

U

W

Y

Made in the USA
Monee, IL
08 November 2026

68020211R00092